Geometry on Poincaré Spaces

Geometry on Poincaré Spaces

by

Jean-Claude Hausmann
and
Pierre Vogel

Mathematical Notes 41

PRINCETON UNIVERSITY PRESS

PRINCETON, NEW JERSEY

1993

Princeton University Press books are printed on acid-free paper and meet the guidelines for permanence and durability of the Committee on Production Guidelines for Book Longevity of the Council on Library Resources

Printed in the United States of America

Library of Congress Cataloging-in-Publication Data

Hausmann, J. C. (Jean-Claude)
Geometry on Poincaré spaces / by Jean-Claude Hausmann and Pierre Vogel.
p. cm.—(Mathematical notes ; 41)
Includes bibliographical references and index.
ISBN 0-691-02113-9
1. Topological spaces. 2. Poincaré series. I. Vogel, Pierre, 1945- . II. Title. III. Series: Mathematical notes (Princeton University Press) ; 41.
QA611.3.H38 1993
514′.3—dc20 92-33615

CONTENTS

Geometry on Poincaré Spaces

1

INTRODUCTION

The theory of manifolds and algebraic topology have a long common history. It is known that the first tools of algebraic topology (the fundamental group, Betti numbers, etc) were forged to distinguish manifolds (See [Di]). But algebraic topology hardly distinguishes spaces of the same homotopy type, while the natural equivalence relation between manifolds is diffeomorphism. The natural question soon arose : how many diffeomorphism classes of manifolds are there in a given homotopy type ? When the homotopy type is that of S^3, this is the famous Poincaré problem.

From this viewpoint it is natural to study the objects of the homotopy category which resemble manifolds the most, namely the spaces satisfying Poincaré duality between homology and cohomology. We call them "Poincaré spaces" (see precise definition in Section 3).

In 1961, W. Browder proved that finite H-spaces satisfy Poincaré duality [Br3]. This is the first occurrence of this notion out of the context of manifolds. Then, in the 1962 Aarhus Topology Symposium [Br4], the same author introduced Poincaré spaces and set what has become the classical question of surgery theory : is there a closed manifold in the homotopy type of a given Poincaré space ? (and how many are there ?)

In the mid-sixties, Poincaré spaces started to be studied in their own right. M. Spivak [Sp] showed that there is an analogue, for a Poincaré space, of the stable normal bundle and the Thom-Pontrjagin class for a closed manifold The difference is that this normal bundle is just a spherical

Serre fibration. (Actually, this spherical fibration together with its Thom-Pontrjagin class is part of our definition of a Poincaré space). C.T.C. Wall, in [Wa4], gives some geometric constructions applicable to Poincaré spaces (cutting and glueing, low-dimensional handles). Examples of Poincaré spaces not homotopy equivalent to a closed topological manifold were produced ([Wa4] and [GS]). Starting with his thesis (Princeton 1967), N. Levitt published several papers on the subject [Le0-4]. Then came the works of L. Jones, F. Quinn, J. Hodgson, J. Lannes-L. Latour-C. Morlet, etc. For a (probably partial, and only dealing with high dimensional Poincaré spaces) list of these works, see, in our reference list, the entries marked by an asterisk.

The interest of this geometry in Poincaré spaces is due to the success of the surgery theory. The power of this theory suggests that the procedure of solving some problems in manifolds (embeddings, group actions, etc) should be divided into two steps : first solve an analogous problem in the category of Poincaré spaces (in which more freedom is expected) and then solve a surgery problem. Among the various manifold constructions one would like to perform directly in the category of Poincaré spaces are :

i) cobordisms
ii) transversality
iii) surgery theory
iv) handle decompositions

Roughly speaking, these constructions were so far obtained by three approaches :

a) Replacing a Poincaré space by a stable thickening of it, which is a manifold with boundary. This point of view is the underlying principle in Levitt's work, as well as in Hodgson's or Lannes-Latour-Morlet' approaches.

b) Endowing Poincaré spaces with "patch space" structures (an atlas whose coordinate changes are homotopy equivalences satisfying appropriate transversality conditions). This is the point of view of L. Jones [Jo1,2].

c) A pure homotopy attack, as tried by F. Quinn [Qu]. No full account was published at the time. For a discussion of Quinn's programm and a partial realization of it, see [Ht].

These three approaches tend to involve a great deal of technical difficulties, of geometrical or homotopical nature, and some of them were not completely overcome. To avoid these problems, a purely homological algebraic conception has been worked out by A. Ranicki [Ra2,3], [LR] and also by A.S. Mishchenko [Mc].

In these notes, we present a new geometric approach to obtain essentially the best current results about the subjects i) to iv) above. Our method is based on standard surgery theory and elementary homotopy arguments, using a cellular approximation of BG starting from BO (see below, the summary of chapter 4). Another ingredient is the fact that a 2-dimensional Poincaré complex is homotopy equivalent to a surface ([E-M] and [E-L]). Otherwise, we do not make use of any result in the subject previously obtained by other authors. These notes are therefore basically self-contained. We only assume some familiarity of the reader with standard surgery theory, as it is exposed in [Wa2, Chapters 1 to 9] and in [Br1].

The chapters are organized as follows :

Chapters 1 and 2 : they contain the definitions leading to the concept of spaces satisfying Poincaré duality. We work in the category of spaces over $\mathbf{R}P^\infty$ (an object in this category is a CW-pair (X,Y) together with a map $w^X : X \longrightarrow \mathbf{R}P^\infty$). Following an idea of Lannes-Latour and Morlet [LLM], we use a homology theory with local coefficients over $\mathbf{R}P^\infty$ which has the advantage of erasing the formal differences between the orientable and non-orientable case (it coincides with ordinary homology with local coefficients when w^X is null-homotopic). The spaces satisfying Poincaré duality between ordinary cohomology with local coefficients and this homology are called **PD-spaces** (the more precise concept of a **Poincaré space**, which can be defined in the category over BG, will be introduced in Chapter 3). In Section 2.C, one finds important results concerning the smoothing of low dimensional skeleta of a PD-space.

Chapter 3 : the notion of a **Q-space** is studied. These are the spaces introduced by F. Quinn [Qu], under the name of "normal spaces" (we changed their name in order to avoid some confusion with the traditional language of surgery). A Q-space of formal dimension n is a space Q over BG (i.e. with a map $j^Q : Q \longrightarrow BG$) equipped with a class (Q) in the n-th homotopy group of the Thom spectrum determined by j^Q. A Poincaré space of formal dimension n is a PD-space and a Q-space of formal dimension n such that the fundamental class $[P] \in H_n(P,\partial P;\mathbf{Z})$ is the image of (P) under a Thom-Hurewicz homomorphism (therefore the map j^P classifies the Spivak bundle of P). Because of the existence and uniqueness theorem for the Spivak bundle, the difference between a PD-space and a Poincaré space is essentially technical; however, the concept of Poincaré space seems to be the proper one for many of our arguments, in which a game is played with various liftings of j^P. At the end of chapter 3, the notions of Poincaré embeddings, Poincaré surgeries and handle subtractions are introduced with some preliminary results, including low dimensional surgeries.

Chapter 4 : It contains the proof of the existence of long exact sequences of the following type :

$$\to {}^{or}\Omega_n^{(PD)}(U) \to H_n(U;MSG) \to L_{n-1}(\pi_1(U)) \to {}^{or}\Omega_{n-1}^{(PD)}(U) \to$$

which replaces the Atiyah-Thom isomorphism $\Omega_n^{SO}(U) \xrightarrow{\cong} H_n(U;MSO)$ when oriented manifolds are replaced by oriented PD-spaces. Such a sequence was found by N. Levitt when $\pi_1(U) = 0$ [Le3], and announced by F. Quinn in the general case [Qu] (it was also found by Jones [Jo1]). For the non-orientable case, we obtain the sequence :

$$\to \Omega_n^{(PD)}(U) \to H_n(U;MG) \to L_{n-1}(\pi_1(U)\times\{\pm 1\};proj_2) \to \Omega_{n-1}^{(PD)}(U).$$

(the relative case is also done). The proof is worked out in the framework of Poincaré spaces : instead of $\Omega_n^{(PD)}$, we consider the Poincaré bordism group $\Omega_n^P(X)$ for a space X over BG (i.e. with a given map $j^X : X \to BG$). This is more general, since ${}^{or}\Omega_n^{(PD)}(U) = \Omega_n^P(U\times BSG)$ and $\Omega_n^{(PD)}(U) = \Omega_n^P(U\times BG)$), where $j^{U\times BSG}$ and $j^{U\times BG}$ are the projections over BG. Poincaré spaces being special cases of Q-spaces one gets a long exact sequence :

$$\to \Omega_n^P(X) \to \Omega_n^Q(X) \to \Omega_n^{QP}(X) \to \Omega_{n-1}^P(X) \to$$

The proof consists of identifying $\Omega_n^Q(X)$ with the homology theory of the Thom spectrum given by j^X (this was done in

Chapter 3) and then identifying $\Omega_n^{QP}(X)$ with a Wall surgery group. We consider successive cellular approximations of $BO \to BG$:

$$BO = B_0 \subset B_1 \subset \cdots \subset \lim B_i = BG$$

Roughly speaking, the proof is by a double induction on the integers n and i for which j^X admits a factorisation through B_i. The case i = 0 reduces to standard surgery theory, with some special arguments in low dimensions (such as the fact that a Poincaré space of formal dimension 2 is homotopy equivalent to a surface [E-M]). The induction step is proven in a linked way with special cases of the $(\pi$-$\pi)$-theorem, surgery and handle subtractions for Poincaré spaces (all obtained by induction on n and i).

The next chapters contain applications of the results or techniques of Chapters 3 and 4.

Chapter 5 : In the first section, we show that a degree-one map $f : P \to R$ between Poincaré spaces of formal dimension n which is covered by a map between the Spivak bundles determines a surgery obstruction $\sigma(f) \in L_n(\pi_1(R))$, and that f is Poincaré bordant to a homotopy equivalence if and only if $\sigma(f) = 0$. In contrast to the case of manifolds, there is no low-dimensional restriction in this statement. We deduce this result in a rather formal way from the results of Chapter 4. A way to obtain these results which is more geometrically direct is given in Section 5.C, by actually "performing surgery", as in manifold theory. This is based on material given in

Section 5.B, in which it is shown how a map from a sphere or a disk to a Poincaré space P can be, in some cases, engulfed into a codimension 0 submanifold of P.

Chapter 6 : It is shown that a Poincaré space P of formal dimension $n \geq 5$ admits Poincaré handle decompositions. In particular, P is a union of compact smooth n-manifolds glued by homotopy equivalences along codimension 0 submanifolds of their boundaries (this is a particular case of a "patch structure" in the sense of L. Jones [Jo1]). An interesting invariant of P is the minimal number $\gamma(P)$ of these manifolds in such a decomposition (see [HV2]).

Chapter 7 : Transversality is one of the main differences between manifolds and Poincaré spaces, and the results of Chapter 4 have for a long time suggested that an obstruction to transversality in Poincaré spaces is to be found in a Wall surgery obstruction group. A large number of results on this subject are proven in this chapter, both for the obstruction to transversality up to Poincaré bordism (Section 7.A), and up to homotopy equivalence (Section B).

Chapter 8 : Some results of Chapter 4 are extended to other kinds of Poincaré spaces (simple Poincaré spaces, Poincaré spaces with coefficients and non-finite Poincaré spaces). Except for the case of simple Poincaré spaces, there are more low-dimensional restrictions.

In an appendix, we prove a proposition on fibrations with finite skeleta, which is used in a couple of places throughout these notes. This result is probably well known to specialists but it is not published in the literature.

AKNOWLEDGMENTS

This work started with a seminar held in the university of Geneva in 1980-81. We would like to thank N. Habegger for a very active participation.

The authors are grateful, for useful conversations and for encouragement, to the following colleagues (in alphabetical order) : W. Browder, F. Conolly, I. Hambleton, E. Pedersen, A. Ranicki, L. Siebenmann, L. Taylor, B. Williams.

Finally, we express our thanks to the referee for pointing out a few mistakes in the first versions of these notes and for valuable suggestions.

1. HOMOLOGY THEORIES OVER $\mathbf{R}P^\infty$

The proper category for working with (possibly non-orientable) Poincaré duality spaces is the category of spaces and pairs over $\mathbf{R}P^\infty$, defined in 1.A. The neatest approach for homology and cohomology is the set up of bundle of groups, or local coefficient systems (1.B). Computations are usually performed using the homology with coefficient in a $\mathbf{Z}\pi$-module, which imply the choice of a connected pointed reference space (1.C). Examples are given in 1.D while Section 1.E is devoted to the presentation of cap-products and the Hurewicz map. Finally we discuss, in Section 1.E, some other homology theories over $\mathbf{R}P^\infty$.

A. THE CATEGORY OVER $\mathbf{R}P^\infty$

Let $\mathbf{R}P^\infty$ denote the infinite dimensional projective space. The role of $\mathbf{R}P^\infty$ will be essentially homotopic : it is just a particular choice of an Eilenberg-McLane space with fundamental group equal to $\{\pm 1\}$.

The natural category for working with (possibly non-orientable) Poincaré duality spaces is the category of **pairs over $\mathbf{R}P^\infty$**. An object of this category is a pair (X,Y) consisting of a CW-complex X and a subcomplex Y, together with a map

$w^X : X \longrightarrow \mathbf{R}P^\infty$. (The spaces X and Y are not pointed).

Throughout these notes, unless explicitly mentioned, whenever we say "a pair (X,Y)", it means an object over $\mathbf{R}P^\infty$ (the map w^X being usually not explicitly mentioned). The pair $(X,\emptyset)$ is treated as a single space and referred to as "the space X" (or "the space X over $\mathbf{R}P^\infty$").

A morphism between (X,Y) and (X',Y') is a map $f : (X,Y) \longrightarrow (X'Y')$ such that

$$w^{X'} = w^X \circ f.$$

(The above equality is an equality of maps, not only up to homotopy). The mapping cylinder $\mathcal{M}_f$ of f

$$\mathcal{M}_f = X \times [0,1] \amalg X' / \{(x,1) = f(x)\}$$

is then considered as a space over $\mathbf{R}P^\infty$ by sending (x,t) to $w^X(x)$ and $x' \in X'$ to $w^{X'}(x')$.

If X is a space over $\mathbf{R}P^\infty$, we also denote by w^X the homomorphism from $\pi_1(X)$ to $\{\pm 1\}$ induced by $w^X : X \longrightarrow \mathbf{R}P^\infty$.

B HOMOLOGY WITH LOCAL COEFFICIENTS

(1.1) For an absolute space X (i.e. X is not over $\mathbf{RP}^{\infty}$), we define the singular chain complex $C_*(X;\mathbf{B})$ with local coefficients in a bundle of abelian groups $\mathbf{B}$ as in [Wh, Chapter VI].

Recall that the **fundamental groupoid** $\pi(X)$ of X is the category whose objects are the points of X and the morphisms from x to y are the homotopy classes of path joining y to x. A **bundle of groups** (or **local coefficient system**) over X is a covariant functor $\mathbf{B}$ from $\pi(X)$ to the category of abelian groups.
A singular k-chain with coefficients in $\mathbf{B}$ is a formal sum

$$c = \sum u(\sigma)\cdot\sigma \qquad (\sigma \in S_k(X))$$

where

a) $S_k(X)$ is the usual set of singular k-simplices of X. An element σ of $S_k(X)$ is a map $\sigma : \Delta^k \longrightarrow X$, where Δ^k is the set of all $(t_0,\dots,t_k)$ in $\mathbf{R}^{k+1}$ with $t_i \geq 0$ and $\sum t_i = 1$

b) $u(\sigma) \in \mathbf{B}(\sigma(1,0,\dots0)))$ and $u(\sigma) = 0$ for almost all σ.

The singular k-chains form an abelian group denoted by $C_k(X;\mathbf{B})$. The boundary operator

$$\partial \;:\; C_k(X;\mathbf{B}) \longrightarrow C_{k-1}(X;\mathbf{B})$$

is defined on $u\cdot\sigma$ $(u \in \mathbf{B}(\sigma(1,0,\ldots 0)))$ by

$$\partial(u\cdot\sigma) = \mathbf{B}\gamma_\sigma(u)\,\partial_0\sigma + \sum(-1)^i\, u\cdot\partial_i\sigma$$

and extended linearly. In this formula, $\partial_i\sigma$ is the usual i^{th}-face of σ and $\gamma_\sigma : [0,1] \longrightarrow X$ is the path $t \longmapsto \sigma(1-t,t,0,\ldots,0)$ (see [Wh, p. 266]).

The homology of the complex $C_*(X;\mathbf{B})$ is denoted by $\dot{H}_*(X;\mathbf{B})$ and its cohomology by $\dot{H}^*(X;\mathbf{B})$. The reason for the notation $\dot{H}_*$ is that the notation H_* is reserved for the homology in the category over $\mathbf{RP}^\infty$, defined in (1.3) below (for the cohomology, we shall see that such a distinction is irrelevant).

If $Y \subset X$, then $C_*(Y;\mathbf{B})$ is a subcomplex of $C_*(X;\mathbf{B})$ and the homology or cohomology of the pair (X,Y) is defined as the homology or cohomology of the quotient complex $C_*(X,Y;\mathbf{B}) = C_*(X;\mathbf{B})/C_*(Y;\mathbf{B})$.

(1.2) Our main examples of local coefficient systems are :

a) If A is an abelian group, the constant bundle $\mathbf{B}_A$ is defined by $\mathbf{B}_A(x) = A$ et $\mathbf{B}_A\gamma = id_A$, for all $x \in X$ and all path γ. We just denote by $\mathbf{Z}$ the constant bundle $\mathbf{B}_\mathbf{Z}$.

b) Let $E \longrightarrow L$ be a real vector bundle of rank r over the space L. Denote by E_z the fiber over $z \in L$. The orientation bundle $\mathbf{Or}_E$ of E is the local system over L defined by $\mathbf{Or}_E(z) =$

$H_r(E_z, E_z - \{0\}; \mathbf{Z})$ which is infinite cyclic; as E is locally trivial, an element of $\mathbf{Or}_E(z)$ can be transported along a path γ^{-1} from z to z', giving the isomorphism $\mathbf{Or}_E\gamma$. We denote just by **Or** the orientation bundle of the canonical line bundle over $\mathbf{R}P^\infty$. Thus **Or** is a local system over $\mathbf{R}P^\infty$.

c) if **B** is a local coefficient system over X and if $f : Y \longrightarrow X$ is a continuous map, the local coefficient system $f^*\mathbf{B}$ over Y is defined by $f^*\mathbf{B}(y) = \mathbf{B}(f(y))$ and $f^*\mathbf{B}\gamma = \mathbf{B}(f \circ \gamma)$. The definition of the tensor product (over **Z**) of two local coefficient systems over X is obvious.

(1.3) Now let (X,Y) be a pair over $\mathbf{R}P^\infty$. Let **B** be a local coefficient system over X. We define $H_*(X,Y;\mathbf{B})$ as the homology of the complex

$$C_*(X,Y;\mathbf{B} \otimes (w^X)^*\mathbf{Or}).$$

As for $H^*(X,Y;\mathbf{B})$, it is the homology of the co-complex

$$\mathrm{Hom}_{\mathbf{Z}}(C_*(X,Y);\mathbf{B})$$

which is just the cohomology of the absolute (i.e. not over $\mathbf{R}P^\infty$) pair (X,Y) [Wh, p. 270].

C. HOMOLOGY THEORIES OVER RP^∞ WITH COEFFICIENTS IN A MODULE

Let (X,Y) be a pair over $\mathbf{RP}^\infty$. Let E be a connected space over RP^∞, with a based point $* \in E$, and let $w_E^X : X \longrightarrow E$ a map over RP^∞. Let $\pi = \pi_1(X)$. These data will enable us to define the homology $H_*(X,Y;B)$ and the cohomology groups $H^*(X,Y;B)$ for a (left-) $Z\pi$-module B. The space E (together with the map w_E^X) is called the **reference space**.

Let $\tilde{E} \longrightarrow E$ be the universal covering of E and consider the following pull-back diagram of regular coverings

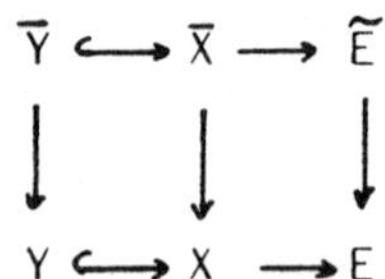

The fundamental group π of E acts by covering transformations on $\tilde{E}$, $\bar{X}$ and $\bar{Y}$, and this action is viewed on the left. Therefore, the usual singular chain complex $C_*(\bar{X};\bar{Y})$ is a complex of $Z\pi$-modules.

All our modules are left-modules, unless explicit mention is made to the contrary. But $Z\pi$ has an anti-involution $\alpha \longmapsto \bar{\alpha}$, such that $\bar{\alpha} = w^E(g)\cdot g^{-1}$ if $\alpha = g \in \pi$. This enables us, as usual, to endow each $Z\pi$-module A with a right $Z\pi$-module structure, denoted by $\bar{A}$, using the convention

$$x\cdot\alpha = \bar{\alpha}\cdot x , \quad x \in A \text{ and } \alpha \quad Z\pi.$$

We define $H_k(X,Y;B)$ as the kth-homology group of the chain complex :

$$\overline{C_*(\bar{X},\bar{Y})} \otimes_{Z\pi} B$$

and $H^k(X,Y;B)$ as the kth-homology group of the cochain complex :

$$\mathrm{Hom}_{Z\pi}(C_*(\bar{X},\bar{Y});B)$$

endowed with the coboundary operator $\beta(z) = (-1)^r\beta(\partial(z))$, for $z \in C_r(X,Y)$, where ∂ is the usual boundary operator of $C_*(\bar{X},\bar{Y})$.

Remarks : 1) If $f : E \to E'$ is a map over $\mathbf{RP}^\infty$ preserving base points, there is a commutative diagram

$$\begin{array}{ccccc} \bar{Y} & \longrightarrow & \bar{X} & \longrightarrow & \tilde{E} \\ \downarrow & & \downarrow & & \downarrow \\ \bar{Y}' & \longrightarrow & \bar{X}' & \longrightarrow & \tilde{E}' \end{array}$$

where the vertical maps are $(\pi_1(E),\pi_1(E'))$-equivariant. Therefore, one has

$$\overline{C_*(\bar{X}',\bar{Y}')} = \overline{C_*(\bar{X},\bar{Y})} \otimes_{Z\pi_1(E)} Z\pi_1(E').$$

In particular, if f induces an isomorphism on the fundamental groups, the homology and cohomology groups defined using E or E' as a reference space are canonically isomorphic.

2) When X is connected, the **standard choice** is E = X, with some fixed base point, and $w_E^X = id_X$. In such a case, the reference space E will not be mentioned (the context is usually clear). Observe that E = X is then used for the homology and cohomology groups of the pairs (X,Y), for all subspace Y of X, as well as for the homology and cohomology groups of these Y's (which is useful since Y is usually not connected in the examples). In particular, one has the natural homomorphism

$$H_k(X,Y;B) \longrightarrow H_{k-1}(Y;B)$$

etc.

2) For a fixed $Z\pi_1(E)$-module B, the cohomology and homology defined above are homology theories over E in the sense of [Do] : they satisfy the Eilenberg-Steenrod axioms but are only defined for pairs over E.

3) The map w^E plays no role in the definition of $H^*(X,Y;B)$. If X is connected (and E = X), the group $H^*(X,Y;B)$ is just the cohomology with local coefficients in B.

4) In contrast, the role of w^E is esential in the definition of $H_*(X,Y;B)$. Even if X is connected, it does not coincide with the homology of (X,Y) with coefficients in B, which is denoted by $\mathring{H}_*(X,Y;B)$. More precisely, if Z denotes, as will be always the case, the additive group of the integers with

trivial π-action, and $\mathbf{Z}^w$ with the twisted action using w^X, then

$$H_*(X,Y;B) = \dot{H}_*(X,Y;Z^w \otimes_Z B).$$

(1.4) The relationship with the homology with local coefficients, defined in Section 1.B, is as follows. Let $\mathbf{B}$ be a local coefficient system over E. Then $B = \mathbf{B}(*)$ is a $\mathbf{Z}\pi$-module. Observe that a simplex $\tilde{\sigma} \in S_k(\tilde{X})$ provides a pair (σ,γ_σ), where $\sigma \in S_k(X)$ is the image of $\tilde{\sigma}$ in X and γ_σ is a path joining the base point * in E to the image of $\sigma(1,0,\ldots,0)$. We can then define a homomorphism of $\mathbf{Z}\pi$-modules

$$\overline{C_*(\tilde{X}) \otimes_{\mathbf{Z}\pi} B} \longrightarrow C_*(X;(w_E^W)^*\mathbf{B} \otimes (w^X)^*\mathbf{Or})$$

by

$$\sigma \otimes b \longmapsto \mathbf{B}(\gamma_\sigma^{-1})(b)\,\sigma$$

As in [Wh, pp 278-281], one proves that this is a chain map which induces an isomorphism on homology. The correspnding treatment for cohomology is similar and left to the reader. Therefore, one has

(1.5) **Proposition** Let (X,Y) be a pair over $\mathbf{R}P^\infty$ and let w_E^X be a reference space for X. Let $\mathbf{B}$ be a local coefficient system over E, with associated π-module B. Then, there are natural isomorphisms :

$$H_*(X,Y;B) \xrightarrow{\simeq} H_*(X,Y;(w_E^X)^*\mathbf{B})$$

$$H^*(X,Y;(w_E^X)^*\mathbf{B}) \xrightarrow{\simeq} H^*(X,Y;B).$$

Proposition (1.5) permits us to compute the homology with local coefficient systems using the homology with coefficients in a module.

D. EXAMPLES

1) If X is connected then

$$H_0(X;B) = B/\{w^X(g)gx - x \mid x \in B,\ g \in \pi_1(X)\}.$$

In particular

$$H_0(X;\mathbf{Z}) = \begin{cases} \mathbf{Z} & \text{if } w^X \equiv 1 \\ \mathbf{Z}_2 & \text{if } w^X \not\equiv 1 \end{cases}$$

2) If X is connected, with standard choice of the reference space, then

$$H_*(X,Y;\mathbf{Z}\pi) = H_*(\tilde{X},\tilde{Y};\mathbf{Z})$$

where $\tilde{X}$ is the universal covering of X and Y the induced covering of Y.

3) $H_{2k}(\mathbf{RP}^\infty;\mathbf{Z}) = \mathbf{Z}_2$ and $H_{2k+1}(\mathbf{RP}^\infty;\mathbf{Z}) = 0$ if $k \geqslant 0$ (standard choice of the reference space).

4) Poincaré-Lefschetz duality holds for a compact manifold M, using our homology and cohomology, provided w^M represents the orientation character of M (see, for instance [Wa2, Theorem 2.1] and [St]). In particular, if M is connected of dimension n, then

$$H_n(M,\partial M;\mathbf{Z}) = H^0(M;\mathbf{Z}) = \mathbf{Z}$$

and there is no formal difference between the orientable case (w^M trivial) and the non-orientable one. Each compact manifold M will be always supposed to be over $\mathbf{R}P^\infty$ and equipped with a generator [M] of $H_n(M,\partial M;\mathbf{Z})$ called its **fundamental class.**

5) If X is connected and simply connected, and $x \in X$, then w^X is homotopic to a constant map. Moreover, for all subspaces Y of X and all local coefficient systems $\mathbf{B}$, the homology groups $H_*(X,Y;\mathbf{B})$ and $\dot{H}_*(X,Y;\mathbf{B})$ are isomorphic to the usual singular homology groups $H_*(X,Y;\mathbf{B}(x))$. But the chain complexes for $H_*(X,Y;\mathbf{B})$ and $\dot{H}_*(X,Y;\mathbf{B})$ are different and give rise to distinct homology theories. For instance, let I = [a,b] with w^I being the constant map to $* \in \mathbf{R}P^\infty$, while I* denotes the same interval with w^{I*} being a loop representing the non-trivial element in $\pi_1(\mathbf{R}P^\infty)$. Observe that $\partial I = \partial I^*$ = {a,b} is the same space over $\mathbf{R}P^\infty$. Then

$$H_0(\partial I;\mathbf{Z}) = H_0(\partial I;\mathbf{Z}) = \mathbf{Z}a \oplus \mathbf{Z}b$$

$$H_1(I,\partial I;\mathbf{Z}) = \mathbf{Z}i, \quad H_1(I,\partial I;\mathbf{Z}) = \mathbf{Z}i^* \quad \text{(chosing a as base point)}$$

but

$$\partial(i) = b - a \qquad \text{while} \qquad \partial(i^*) = -b - a.$$

6) Let $S^* = S^1$ be the circle, considered over $\mathbf{R}P^\infty$ so that w^{S^*} is not homotopic to a constant map. As

$$H_1(S^*;\mathbf{Z}) = \dot{H}_1(S^1;\mathbf{Z}^w) = \dot{H}^0(S^1;\mathbf{Z}^w) = 0,$$

one has

$$H_i(S^*;Z) = \begin{cases} Z_2 & \text{if } i = 0 \\ 0 & \text{otherwise.} \end{cases}$$

E. CAP-PRODUCTS - THE HUREWICZ HOMOMORPHISM

We shall use the cap-product :

$$H^k(X,Y_1;B_1) \times H_i(X,Y_1 \cup Y_2;B_2) \longrightarrow H_{i-k}(X,Y_2;B_1 \otimes_Z B_2)$$

where B_i are $Z\pi$-modules ($\pi = \pi_1(X)$) and $B_1 \otimes_Z B_2$ is equipped with the $Z\pi$-module structure determined by $g(b_1 \otimes b_2) = gb_1 \otimes gb_2$. The cap product is induced by the pairing at the chain level

$$\mathrm{Hom}_{Z\pi}(C_k(X,Y_1);B_1) \times \overline{C_i(X,Y_1 \cup Y_2)} \otimes_Z B_2$$

$$\overline{C_{i-k}(X,Y_2)} \otimes_{Z\pi} (B_1 \otimes_Z B_2)$$

$$(\alpha,\sigma \otimes b_2) \longmapsto \sigma^{i-k} \otimes [\alpha({}^k\sigma) \otimes b_2]$$

where ${}^k\sigma$ and σ^{i-k} are the front k-face and the back (i-k)-face of the singular i-simplex σ (see [Sp. p. 250]). The usual properties of cap-products hold. The corresponding treatment may be done with homology with local coefficients, either using Proposition (1.5) or directly.

Let X be a space over $\mathbf{RP}^\infty$, and $x \in X$. The Hurewicz homomorphism $h : \pi_n(X,x) \longrightarrow H_n(X;\mathbf{Z})$ is defined as follows : let $\tilde{\alpha} : S^n \longrightarrow X$ be a representative of $\alpha \in \pi_n(X,x)$. Then the standard sphere $S = S^n$ becomes a space over $\mathbf{RP}^\infty$ by $w^S = w^X \circ \tilde{\alpha}$. If $w^S : \pi_1(S) \longrightarrow \{\pm 1\}$ is trivial (for instance, when $n \geqslant 2$), one defines $h(\alpha) = \tilde{\alpha}_*([S^n])$, where $[S^n]$ is the standard orientation class of S^n, $[S^n] \in H_n(S^n;\mathbf{Z}) = \dot{H}_n(S^n;\mathbf{Z})$. When $S = S^1$ and $w^S \neq 1$, one defines $h(\alpha) = 0$ (recall that $H_1(S;\mathbf{Z}) = 0$, as explained in example 6, Section 1.D). The Hurewicz homomorphism for pairs over $\mathbf{RP}^\infty$, $h : \pi_n(X,Y) \longrightarrow H_n(X,Y;\mathbf{Z})$ is defined in the same way.

In the local coefficient set up, recall that, for an absolute space X, one has a bundle of groups $\boldsymbol{\pi}_n X$ so that $\boldsymbol{\pi}_n X(x) = \pi_n(X,x)$ [Wh, p. 257]. For a space X over $\mathbf{RP}^\infty$, the relevant bundle is $\boldsymbol{\pi}_n X = \boldsymbol{\pi}_n X \otimes w^{X*}\mathbf{Or}$. This gives another $\mathbf{Z}\pi$-module ($\pi = \pi_1(X,x_0)$) on $\pi_n(X,x_0)$, twisted by w^X. For this module structure, the homomorphism $\pi_n(X,x_0) \longrightarrow H_n(\tilde{X};\mathbf{Z}) = H_n(X;\mathbf{Z}\pi)$ is a homomorphism of $\mathbf{Z}\pi$-modules.

F. OTHER HOMOLOGY THEORIES OVER $\mathbf{RP}^\infty$

We shall use a couple of homology theories over $\mathbf{RP}^\infty$ in the sense of [Do] (they satisfy the Eilenberg-Steenrod axioms but are only defined for pairs over $\mathbf{RP}^\infty$). Let h_* be such a homology theory. The groups $h_*(X,Y)$ are the abutment of an Atiyah-Hirzebruch spectral sequence whose E^2-term is the usual homology $\dot{H}_p(X,Y;\underline{h_q(pt)})$ with local coefficient in $\underline{h_q(pt)}$ considered as a local system over X (see [Do, p. 394]). If X is connected, the homology in this local system is the homology with the $Z\pi$-module $\underline{h_q(*)}$, where * is a base point of X. The homology theory h_* is called a **w-homology theory** if, for any connected space X over $\mathbf{RP}^\infty$, the $\pi_1(X)$-module structure on $\underline{h_q(*)}$ is given by $gu = w^X(g)u$, for $g \in \pi_1(X,*)$ and $u \in \underline{h_q(*)}$. In such a case, the module $\bar{Z}\otimes_Z \underline{h_q(*)}$ coincides with $h_*(*)$ endowed with the trivial $\pi_1(X,*)$-action. As $\dot{H}_p(X,Y;\bar{Z}\otimes_Z \underline{h_q(*)}) = H_p(X,Y;h_q(*))$, this proves the following :

(1.7) <u>Proposition</u> Let h_* be a w-homology theory over $\mathbf{RP}^\infty$. Then, for any pair (X,Y) over $\mathbf{RP}^\infty$, there is a spectral sequence

$$E^2_{p,q} = H_p(X,Y;h_q(*)) \Longrightarrow h_{p+q}(X,Y)$$

where $h_q(*)$ has the trivial $\pi_1(X)$-action.

(1.8) **Example** Let $\Omega_n^w(X,Y)$ be the bordism classes of pairs (M,f) where :

1) M is a compact smooth manifold (equipped with its fundamental class $[M] \in H_n(M,\partial M;\mathbb{Z})$ and with its orientation character $w^M : M \longrightarrow \mathbf{R}P^\infty$; we sometimes use the notation $(M,[M],f)$).

2) $f : (M,\partial M) \longrightarrow (X,Y)$ is a map over $\mathbf{R}P^\infty$.

Observe that $\Omega_n^w(*)$ is the oriented bordism of the point. Let us show that Ω_n^w is a w-homology theory. The fact that the bordism theory satisfies the homology axioms is classical. Let $\gamma : [0,1] \longrightarrow X$ be a loop at $*$. Let (M,const.) represent an element of $\Omega_n^w(*)$. Consider $M\times[0,1]$ as a space over $\mathbf{R}P^\infty$ by $w = w^{M\times[0,1]} = w^X \circ g$, where $g : M\times[0,1] \longrightarrow X$ is defined by $g(x,t) = \gamma(t)$. According to [Do, 2.4], the image of $(M,[M],\mathrm{const})$ under the action of $\gamma \in \pi_1(X,*)$ is equal to $(M,[M]',\mathrm{const})$, where $[M]'$ is the image of $[M]$ under the homomorphism

$$H_n(M,\partial M;\mathbb{Z}) \xrightarrow{i_0} H_n(M\times[0,1],\partial M\times[0,1];\mathbb{Z}) \xrightarrow{i_1^{-1}} H_n(M,\partial M;\mathbb{Z})$$

where i_t is induced by the inclusion $M \simeq M\times\{t\} \hookrightarrow M\times[0,1]$. Consider $\mathbb{Z}$ as the trivial local system over $M\times[0,1]$. Let a be an n-cycle of the complex $C_*(M\times\{0\},\partial M\times\{0\};\mathbb{Z})$ representing $[M]$. Viewed as an element of $C_n(M\times[0,1],\partial M\times[0,1];\mathbb{Z})$, the cycle a is homologous to the element $w(\gamma)a$ in $C_n(M\times\{1\},\partial M\times\{1\};\mathbb{Z})$. Therefore, $[M]' = w(\gamma)[M]$ and Ω_*^w is a w-homology theory.

Other examples will appear in Section 3.

2. POINCARE DUALITY - BASIC PROPERTIES

A. PD-SPACES

Among the many possible definitions of spaces satisfying Poincaré duality, we shall work, most of the time, with the two given in (2.2) below. We present these two definitions as particular cases of a more general concept. We use the language of homology with coefficients in modules. We could instead use homology with local coefficients, as seen in Section 1.

(2.1) DEFINITION OF A PD-SPACE. Let P be a connected space over $\mathbf{R}P^{\infty}$. Let ∂P be a subspace of P with finitely many connected components $\partial_i P$ ($i = 1,2,\ldots,k$). Let $\Lambda_1,\Lambda_2,\ldots,\Lambda_k$ and Λ be rings with anti-involution (denoted by $x \mapsto \bar{x}$) which fit in a commutative diagram of homomorphisms of rings with anti-involution :

$$\begin{array}{ccc} \mathbf{Z}\pi_1(\pi_i P) & \longrightarrow & \Lambda_i \\ \downarrow & & \downarrow \\ \mathbf{Z}\pi_1(P) & \longrightarrow & \Lambda \end{array}$$

Write $\Lambda_\partial = \coprod \Lambda_i$. We say that P is a **PD-space over** $(\Lambda,\Lambda_\partial)$, of **formal dimension** n, with **boundary** ∂P, if P is equipped with a class $[P] \in H_n(P,\partial P;\mathbf{Z})$, called the **fundamental class** of P, such that :

1) P and ∂P are homotopy equivalent to finite complexes

2) For all j and for any Λ-module B, the cap product

$$- \cap [P] \; : \; H^j(P;B) \longrightarrow H_{n-j}(P,\partial P;B)$$

is an isomorphism.

3) Let $[\partial_i P]$ be the ith component of the image of $[P]$ under the homomorphism

$$H_n(P,\partial P;\mathbf{Z}) \longrightarrow H_{n-1}(\partial P;\mathbf{Z}) = \bigoplus_j \; H_{n-1}(\partial_j P;\mathbf{Z}).$$

Then

$$- \cap [\partial_i P] \; : \; H^j(\partial_i P;B) \longrightarrow H_{n-j}(\pi_i P;B)$$

is an isomorphism for all j and all Λ_i-module B.

This definition extends to the non-connected case by applying it component-wise (with the same formal dimension). The case $\partial P = \emptyset$ is permitted.

(2.2) REMARKS and EXAMPLES :

1) Condition 1 in Definition (2.1) is taken for simplicity. For our purpose, we could use the following more general condition

1') : P and ∂P have the homotopy type of a complex with finite skeleta and the Wall finiteness obstructions of P and ∂P which can be defined in $K_0(\Lambda)$ and $K_0(\Lambda_i)$

(because of Conditions 2) and 3)), are all equal to 0.

2) Definition (2.1) will be mainly used in the following two cases :

a) the case $\Lambda_i = \Lambda$ for all i. The space P is then called a **(Λ-PD)-space**. The obvious generalization of the argument of [Br1, Section 2] shows that Condition 3 follows from Condition 2.

b) the case $\Lambda = Z\pi_1(P)$ and $\Lambda_i = Z\pi_1(\partial_i P)$. We say that P is a **PD-space**. Observe that a $(Z\pi_1(P))$-PD-space is not in general a PD-space, unless $\pi_1(\partial_i P)$ injects in $\pi_1(P)$. Conditions 1) and 1') are equivalent for PD-spaces of formal dimension ≥ 3, using [Bn, Section 3] and [Wa 3].

3) Observe that a PD-space is realy a couple (P,[P]), the fundamental class [P] being part of the definition (like in the concept of an oriented manifold). When there is no danger of confusion, we use the shorter notations P for (P,[P]) and -P for (P,-[P]). If P is a (Z-PD)-space, (P,[P]) and (P,-[P]) are the only possible (Z-PD)-spaces with underlying space P. This is not true in general : for instance, if P is a (Z-PD)-space of formal dimension n, and if [[P]] is any non-zero class in $H_n(P,\partial P;Z)$, then (P,[[P]]) is a (**Q**-PD)-space.

A continuous map $f : P \to P'$ between PD-spaces (or Λ-PD-spaces) is called a **morphism of PD-spaces** (or Λ-PD-spaces) if $f_*([P]) = [P']$.

The letters "PD" stand of course for "Poincaré Duality". However, we draw the attention of the reader to the distinction between a PD-space, which is a space over $\mathbf{R}P^\infty$, and a "Poincaré space", a concept in the category over BG which is to be defined in Section 3. But for the time being, note that our concept of a PD-space coincides with those of [Wa 2], except that we do not worry about simplicity. Our formalism and definitions erase the formal distinction between the orientable and the non-orientable case. Therefore, though we might keep calling $w^P : \pi_1(P) \longrightarrow \{\pm 1\}$ the **orientation character** of P and saying that P is **orientable** if $w^P \equiv 1$, there will be almost no need for these names throughout these notes.

(2.3) PD-COBORDISMS Let P_i $(i = 1,2)$ be two PD-spaces, with $\partial P_1 = \partial P_2$. A PD-space W such that $\partial W = P_1 \cup \partial P_1 \times [0,1] \cup P_2$ is called a **PD-cobordism** between P_1 and P_2. The cap-product with [W] gives then an isomorphism $H^k(W,P_1;B) \xrightarrow{\cong} H_{n+1-k}(W,P_2;B)$ (n = formal dimension of P_i), for any $Z\pi_1(P)$-module B. One can define the notion of a (Λ-PD)-cobordism in the same way, for a homorphism of rings with anti-involution $Z\pi_1(P) \longrightarrow \Lambda$.

<u>Examples</u> :

1) Let P be a PD-space of formal dimension n. We call $P\times[0,1]$ the PD-space $(Q,[Q])$ of formal dimension n+1 defined as follows : as a space over $\mathbf{RP}^\infty$, one sets $Q = P\times[0,1]$ and $w^Q(x,t) = w^P(x)$. The boundary ∂Q is defined as

$$\partial Q = P\times\{0\} \cup P\times[0,1] \cup P\times\{1\}$$

As $H_*(P\times[0,1],P\times\{i\};\mathbb{Z}) = 0$ for all $i \in [0,1]$, the long exact sequence of the triple $(Q,\partial Q,P\times\{1\})$ gives the isomorphism

$$H_{n+1}(Q,\partial Q;\mathbb{Z}) \xrightarrow{\cong} H_n(\partial Q,P\times\{1\};\mathbb{Z}) \cong H_n(P\times\{0\},\partial(P\times\{0\});\mathbb{Z})$$

One defines $[Q]$ to be the class going to $-[P]$ under this isomorphism. With this definition, $P\times[0,1]$ is a PD-cobordism between P and itself.

2) A non-trivial extension of w^P permits us to see $P\times[0,1]$ as a PD-cobordism between P and -P (see the proof of (2.6)).

3) Let P_i $(i = 1,2)$ be $(\Lambda\text{-PD})$-spaces, with $\partial P_1 = \partial P_2$. Let $f : P_1 \longrightarrow P_2$ be a Λ-homotopy equivalence (over $\mathbf{RP}^\infty$) such that $f|\partial P_1 = \mathrm{id}$ and so that the following diagram

$$\begin{array}{ccc} \mathbb{Z}\pi_1(P_1) & \longrightarrow & \mathbb{Z}\pi_1(P_2) \\ & \searrow \quad \swarrow & \\ & \Lambda & \end{array}$$

commutes. We suppose also that $f_*([P_1]) = \pm[P_2]$. Then the mapping cylinder $\mathcal{M}_f$ of f, considered as a space over $\mathbf{RP}^\infty$ as in 1.A, is a $(\Lambda\text{-PD})$-cobordism between P_1 and $\pm P_2$.

B. PD-DECOMPOSITIONS

Let P be a space over $\mathbf{RP}^\infty$, together with a subspace ∂P and a class $[P] \in H_n(P,\partial P;Z)$. Let $Z\pi_1(P) \longrightarrow \Lambda$ be a homomorphism of rings with anti-involution. Let us consider a decomposition into subcomplexes :

$$P = P_- \cup P_+ \ , \quad P_- \cap P_+ = P_0 \ , \quad P_\pm \cap \partial P = \partial_\pm P.$$

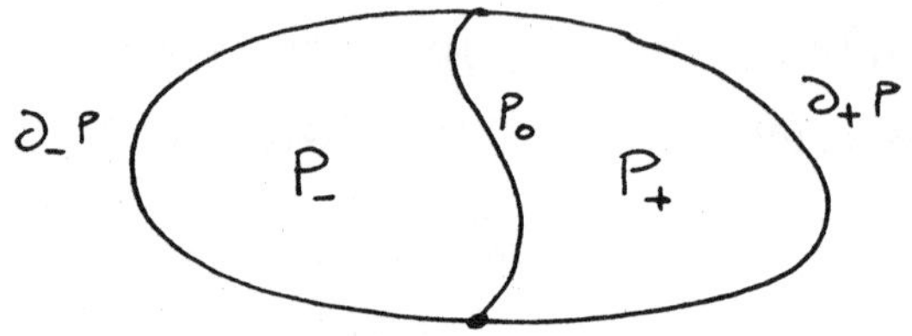

This is called a **PD-decomposition** (respectively, a **(Λ-PD)-decomposition**) if

1) all the spaces $\partial P_\pm$, P_0, and $\partial_\pm P$ are PD-spaces (respectively, (Λ-PD)-spaces) with

$$\partial P_\pm = P_0 \cup \partial_\pm P \ , \quad \partial P_0 = P_0 \cap \partial P \quad \text{and} \quad \partial(\partial_\pm P) = \partial P_0.$$

2) the image of [P] under the homomorphisms

$$H_n(P,\partial P;Z) \longrightarrow H_n(P,\partial P \cup P_+;Z) \longrightarrow H_n(P_-,\partial P_-;Z) \quad \text{and}$$

$$H_n(P,\partial P;Z) \longrightarrow H_n(P,\partial P \cup P_-;Z) \longrightarrow H_n(P_+,\partial P_+;Z)$$

is $[P_\pm]$.

3) P_- is a PD-cobordism between $\partial_- P$ and P_0 and P_+ is a PD-cobordism between P_0 and $\partial_+ P$ (respectively, (Λ-PD)-cobordisms).

Observe that $\partial P = \partial_- P \cup \partial_+ P$ is a PD-decomposition (respectively, a (Λ-PD)-decomposition) of ∂P.

PD-decompositions occur in the operations of "pasting and cutting" which are describe below.

(2.4) PASTING PD-SPACES. Let P_- and P_+ be two PD-spaces (respectively, (Λ-PD)-spaces) of formal dimension n. Suppose that the boundaries $\partial P_{\pm}$ are equipped with PD-decompositions (respectively, (Λ-PD)-decompositions) : $\partial(P_{\pm}) = \partial_{\pm} P \cup P_0^{\pm}$. Let $h : P_0^+ \to P_0^-$ be a homotopy equivalence (respectively, a Λ-homology equivalence such that its mapping cylinder $\mathcal{M}_h$ is homotopy equivalent to a finite complex), over $\mathbf{R}P^\infty$. The space

$$P = P_- \cup \bar{P}_+,$$

where $\bar{P}_+ = P_+ \cup \mathcal{M}_h$, is naturally a space over $\mathbf{R}P^\infty$, (see 1.A). Observe that $\bar{P}_+$ is a PD-space with fundamental class $[\bar{P}_+]$ corresponding to $[P_+]$ under the isomorphism $H_n(\bar{P}_+, \partial\bar{P}_+) = H_n(\bar{P}_+, \partial\bar{P}_+ \cup \mathcal{M}_h) = H_n(P_+, \partial P_+)$.

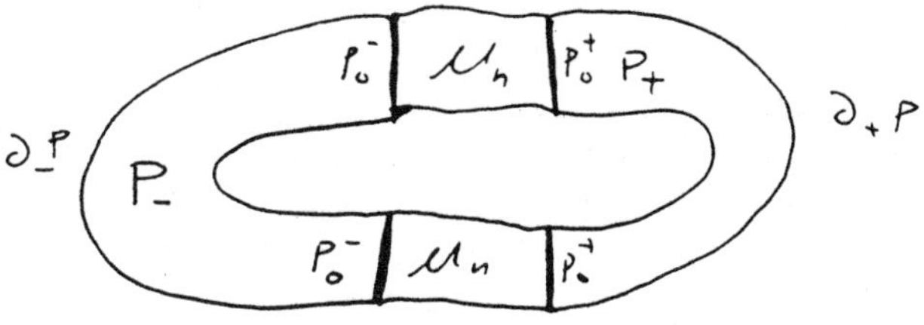

(2.5) **Proposition** Suppose that $h_*([P_0^+]) = -[P_0^-]$. Then there is a class $[P] \in H_n(P,\partial P;\mathbb{Z})$ such that P is a PD-space (respectively, (Λ-PD)-space) with fundamental class [P], and P $= \overline{P}_+ \cup P_-$, with $\overline{P}_+ = P_+ \cup \mathcal{U}_h$ is a PD-decomposition (respectively, a (Λ-PD)-decomposition).

Proof : The homology (with $\mathbb{Z}$ coefficients) exact sequence of the triple $(P,\partial P \cup P_0^-,\partial P)$ gives the following commutative diagram :

$$\begin{array}{ccccc} H_n(P;\partial P) \longrightarrow & H_n(P;\partial P \cup P_0^-) & \longrightarrow & H_{n-1}(\partial P \cup P_0^-,\partial P) \\ & \wr\wr & & \wr\wr \\ & H_n(P_+,\partial P_+) \oplus H_n(P_-,\partial P_-) & \overset{\beta}{- - \rightarrow} & H_{n-1}(P_0^-,\partial P_0^-) \end{array}$$

As $h_*([P_0^+]) = -[P_0^-]$, one has $\beta([P_+],[P_-]) = 0$, and therefore there is a class [P] $H_n(P,\partial P)$ so that $\alpha([P]) = ([P_+],[P_-])$. For any $Z\pi_1(P)$-module (respectively, any Λ-module) B, there is a ($\pm$)-commutative diagram, which is displayed on the next page (Diagram (*)). By the five Lemma, the homomorphism $-\cap[P]$: $H^k(P;B) \longrightarrow H_{n-k}(P,\partial P;B)$ is an isomorphism. The finiteness condition 1) of Definition (2.8) is easily verified. Therefore P is a PD-space (respectively, a (Λ-PD)-space) and Proposition (2.11) is proven. []

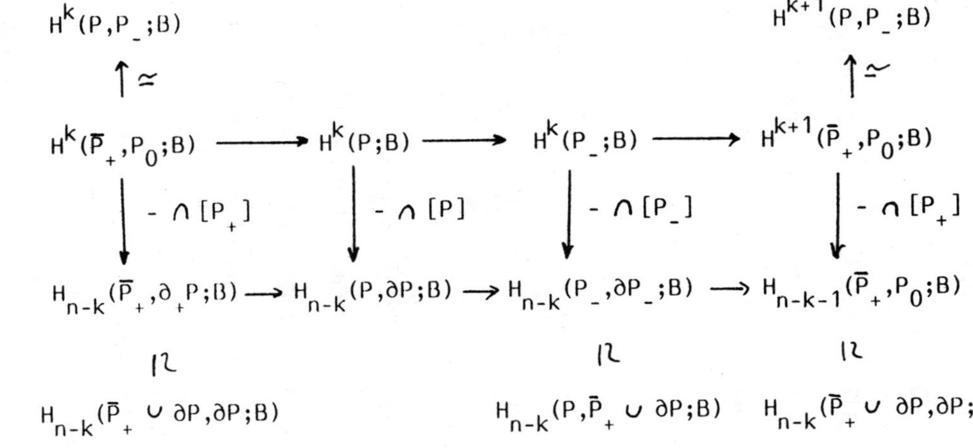
H^k(P,P_;B)
≃
H^{k+1}(P,P_;B)
≃
H^k(P̄_+,P_0;B)
H^k(P;B)
H^k(P_;B)
H^{k+1}(P̄_+,P_0;B)
- ∩ [P_+]
- ∩ [P]
- ∩ [P_]
- ∩ [P_+]
H_{n-k}(P̄_+,∂_+P;B)
H_{n-k}(P,∂P;B)
H_{n-k}(P_,∂P_;B)
H_{n-k-1}(P̄_+,P_0;B)
≀≀
≀≀
≀≀
H_{n-k}(P̄_+ ∪ ∂P,∂P;B)
H_{n-k}(P,P̄_+ ∪ ∂P;B)
H_{n-k}(P̄_+ ∪ ∂P,∂P;B)

Diagram (*)

Remark The class [P] of Proposition(2.5) is not unique in general : two such classes differ by an element of the image of $H_n(P_0^-,\partial P_0^-) \longrightarrow H_n(P,\partial P)$. Observe that this image is zero in the case of $\mathbb{Z}$-Poincaré spaces.

The hypothesis that $h_*([P_0^+]) = -[P_0^-]$, of Proposition (2.5), might not be fulfilled in practice, as for instance when attaching a non-orientable handle of index 1 to an orientable manifold. In general,

$$H_{n-1}(P_0^{\pm},\partial P_0^{\pm}) = \bigoplus_i H_{n-1}(T_i^{\pm},\partial T_i^{\pm})$$

where $T_i^{\pm}$ are the connected components of $P_0^{\pm}$. Here is a version of Proposition (2.5) under the weaker hypothesis $h_*([T_i^+]) = \pm[T_i^-])$.

(2.6) **Proposition** Suppose that $h_*([T_i^+]) = \varepsilon_i \cdot [T_i^-])$, where $\varepsilon_i = \pm 1$. Then, there is a canonical extension $w^P : P \longrightarrow \mathbf{R}P^\infty$ of w^{P_+} and w^{P_-} and a class $[P] \in H_n(P,\partial P;\mathbb{Z})$ such that P is a PD-space (respectively, $(\Lambda$-PD)-space) with fundamental class [P], and $P = \overline{P}_+ \cup P_-$, with $\overline{P}_+ = P_+ \cup \mathcal{M}_h$ is a PD-decomposition (respectively, a $(\Lambda$-PD)-decomposition).

The proof of (2.6) uses the following property of $\mathbf{R}P^\infty$:

(2.7) **Lemma** There exist a map $\mu : \mathbf{R}P^\infty \times [0,1] \longrightarrow \mathbf{R}P^\infty$ such that, for all $x \in \mathbf{R}P^\infty$ one has

1) $\mu(x,0) = \mu(x,1) = x$

2) μ $\{x\} \times [0,1]$ represents a non-trivial loop in $\mathbf{RP}^\infty$.

Proof : Let $p : S^\infty \longrightarrow \mathbf{RP}^\infty$ be the universal covering of $\mathbf{RP}^\infty$. It is equivalent to define a map $\mu : S^\infty \times [0,1] \longrightarrow S^\infty$ such that $\mu(x,0) = x$, $\mu(x,1) = -x$ and $p(\mu(x,\tau)) = p(\mu(-x,\tau))$.

The space S^∞ is homeomorphic to the infinite join $S^1 * S^1 * \ldots$. An element $x \in S^\infty$ can then be represented by a sequence $(z_j, t_j)_{j \in \mathbf{N}}$ where z_j is a complex number of modulus 1, $t_j \in [0,1]$, $t_j = 0$ for almost all j and $\sum t_j = 1$. One defines

$$\mu((z_j, t_j), \tau) = ((e^{i\pi\tau} z_j, t_j) \qquad [].$$

Proof of (2.6) : Let M_i be the mapping cylinder of $h|T_i$. Thus $P_+ = P_+ \cup \bigcup_i M_i$. Define w^{M_i} by

$$w^{M_i}(x,t) = w^{P_+}(x) \qquad \text{if} \quad \mathcal{E}_i = -1$$

$$w^{M_i}(x,t) = \mu(w^{P_+}(x), t) \qquad \text{if} \quad \mathcal{E}_i = 1$$

where μ is the map of Lemma (2.7). We can then use the argement of the proof of Propoition (2.5).

[]

Remark The hypothesis $\mathcal{E}_i \neq \pm 1$ of Proposition (2.6) cannot be weakened, even for $\Lambda = \mathbf{Q}$. For instance, if $P_\pm = S^1 \times [0,1]$, the $S^1 \times \{0\}$'s being glued by the identity and the $S^1 \times \{1\}$'s by a map of degree $\neq 0, \pm 1$, then $H_2(P;\mathbf{Z}) = 0$.

(2.8) CUTTING PD-SPACES : The operation of cutting a PD-space along a codimension 1 PD-space is more canonical than the previous operation of pasting. We describe a few important cases.

(2.9) **Proposition** Let P, ∂P, Λ, and a decomposition of P as above. Suppose that :

a) All the spaces under consideration are homotopy equivalent to finite complexes.

b) P is a (Λ-PD)-space with boundary ∂P.

c) P_- is a (Λ-PD)-space and $\partial P_- = \partial_- P \cup P_0$ is a (Λ-PD)-decomposition. (Here, $[P_-]$ is the image of $[P]$ under the usual homomorphisms.)

Let $[P_+]$ be the image of $[P]$ under the homomorphism $H_n(P,\partial P;\mathbf{Z}) \longrightarrow H_n(P,\partial P \cup P_-;\mathbf{Z}) \cong H_n(P_+,\partial P_+;\mathbf{Z})$. Then, P_+ is a (Λ-PD)-space with fundamental class $[P_+]$, boundary $\partial P_+ = \partial_+ P \cup P_0$, and the above decomposition is a (Λ-PD)-decomposition of P.

<u>Proof</u> : Poincaré duality for P_+ is proved as in the proof of Proposition (2.6). [].

For PD-spaces, we prove in the same way :

(2.10) **Proposition** Let P, ∂P, and a decomposition of P as above. Suppose that :

a) All the spaces under consideration are homotopy equivalent to finite complexes.

b) P is a PD-space with boundary ∂P. ($[P_-]$ being the image of $[P]$ under the usual homomorphisms.)

c) P_- is a PD-space and $\partial P_- = \partial_- P \cup P_0$ is a PD-decomposition.

d) $\pi_1(P_0) \cong \pi_1(P_-)$ and $\pi_1(\partial P_0) \cong \pi_1(\partial_- P)$.

Let $[P_+]$ be the image of $[P]$ under the homomorphism $H_n(P,\partial P;Z) \longrightarrow H_n(P,\partial P \cup P_-;Z) \cong H_n(P_+,\partial P_+;Z)$. Then, P_+ is a PD-space with fundamental class $[P_+]$, boundary $\partial P_+ = \partial_+ P \cup P_0$, and the above decomposition is a PD-decomposition of P. []

(2.11) POINCARÉ EMBEDDINGS : A special case of PD-decomposition is given by a PD-embedding. Let Z be a Poincaré space of formal dimension k and P be a Poincaré space of formal dimension n. A **PD-embedding** (of codimension n-k) of Z into P consists of :

a) an (n-k-1)-sphere bundle $\nu : E \longrightarrow Z$, (i.e. a Serre fibration with fiber homotopy equivalent to S^{n-k-1}). It has an orientation character $w^{\nu} : Z \longrightarrow \mathbf{R}P^{\infty}$. The mapping cylinder $\hat{E}$ of ν then becomes a PD-space of formal dimension n with $\partial\hat{E} = E$.

b) a PD-space $\overline{P}$ of formal dimension n and a PD-decomposition $\overline{P} = \overline{P}_- \cup \overline{P}_+$, etc, of P, with $(\hat{E}, \mathcal{M}_{\nu|\partial Z}, E) = (\overline{P}_-, \partial_-\overline{P}, \overline{P}_0)$ as triple of PD-spaces.

c) a homotopy equivalence $\alpha : \overline{P} \longrightarrow P$ of degree one.

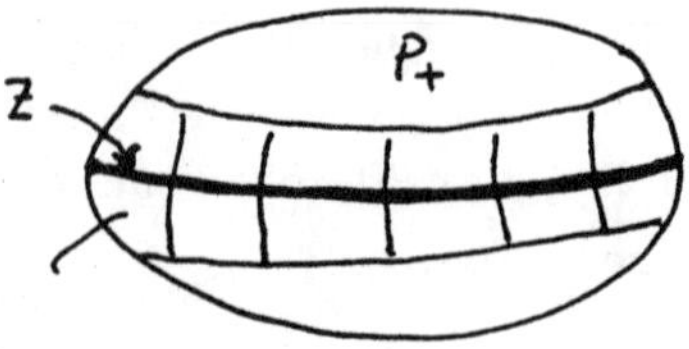

The cases $\partial Z = \emptyset$ and $\partial Z = \partial P = \emptyset$ are of course possible. When $\partial Z = \emptyset$, we always suppose that $\partial\overline{P} = \partial P$ and $\alpha|\partial\overline{P} = \text{id}$. This is possible by gluing to P the mapping cylinder of $\alpha|\partial\overline{P}$, which gives us a new P and a new α satisfying these conditions.

The bundle ν is the **normal bundle** of the PD-embedding. If $f : Z \longrightarrow P$ is a map over $\mathbb{R}P^\infty$ and if there is a PD-embedding as above such that $\alpha|Z = f$, we then say that f is equivalent to a PD-embedding, or that the homotopy class of f is realizable by a PD-embedding. The homotopy equivalence α is often forgotten in the notation and we simply identify $\overline{P}$ with P (and write $P = \hat{E} \cup P_+$, etc).

If Z and P are (Λ-PD)-spaces and α is just a Λ-homology equivalence, we talk about a **(Λ-PD)-embedding** of Z into P.

Remark : This notion of PD-embedding is what is esssentially the concept of a Poincaré embedding introduced by Browder (see also [Wa2]). However, we reserve this name for the corresponding concept in the category of Poincaré spaces which will appear in Section 3.C.

C. SMOOTHING LOW-DIMENSIONAL SKELETA

The results of this section are essential for the next chapters. For instance, they make possible low-dimensional surgeries (of index ≤ 2) in PD-spaces.

Let $r : \Lambda' \longrightarrow \Lambda$ be a homomorphism of rings. We say that r is **locally epic** if for all $a_1,\ldots,a_r \in \Lambda$, there is a unit u of Λ such that $ua_1,\ldots ua_r \in$ Image(r) (see [CS1, p. 288]).

(2.15) **Proposition** Let X be a PD-space (respectively, a (Λ-PD)-space, with $Z\pi_1(X) \longrightarrow \Lambda$ locally epic) of formal dimension $n \geq 5$. Then, there exists

a) a compact manifold W of dimension n, with a manifold decomposition $\partial W = W' \cup W''$ (W' may be empty). The manifold W' admits a handle decomposition with handles of index ≤ 2 (≤ 1 if $n = 5$) and the pair (W,W') admits a handle decomposition with handles of index ≤ 2.

b) a PD-space (respectively, a (Λ-PD)-space) X_2 of formal dimension n with a PD-decomposition (respectively, a (Λ-PD)-decomposition) $\partial X_2 = W'' \cup X_2'$. The pair (X_2,W'') is 2-connected. For the pair $(X_2',\partial W')$, there are the following three possibilities

- W' is empty, or
- $(X_2', \partial W')$ is 2-connected when $n \geqslant 6$, or
- $(X_2', \partial W')$ is 1-connected when $n = 5$

c) a morphism of PD-spaces $F: (Z, \partial Z) \longrightarrow (X, \partial X)$, where Z is the PD-space (respectively, a (Λ-PD)-space) $Z = X_2 \cup_{W''} W$, with boundary $\partial Z = X_2' \cup_{\partial W'} W'$. The maps F and $F|\partial Z$ are homotopy equivalences (respectively, maps inducing an isomorphism on the fundamental groups and on the homology with Λ as coefficients)

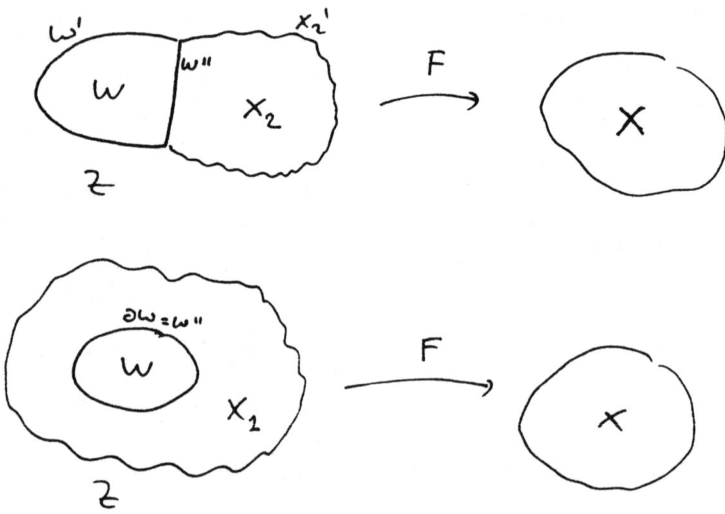

The manifold W is called a **smooth 2-skeleton** of $(X, \partial X)$ (or of X if W' is empty). When $W' \neq \emptyset$, then W' is called the smooth 2-skeleton (or 1-skeleton if $n = 5$) of ∂X **induced** by W.

PD-spaces of formal dimensions 3 and 4 have smooth 0 and 1 skeleta respectively. The precise results are :

(2.16) **Proposition** Let X be a PD-space (respectively, a (Λ-PD)-space, with $Z\pi_1(X) \longrightarrow \Lambda$ locally epic) of formal dimension 4, with $\partial X = \emptyset$. Then, there exists

a) a compact manifold W of dimension 4 admiting a handle decomposition with handles of index ≤ 1.

b) a PD-space (respectively, a (Λ-PD)-space) X_1 of formal dimension 4 such that $\partial X_1 = \partial W$. The pair $(X_1, \partial W)$ is 1-connected.

c) a morphism of PD-spaces $F : Z \longrightarrow X$, where Z be the PD-space (respectively, the (Λ-PD)-space) $Z = X_1 \cup_{\partial W} W$. The map F is a homotopy equivalence (respectively, a map inducing an isomorphism on the fundamental groups and on the homology with Λ as coefficients)

(2.17) **Proposition** Let X be a PD-space of formal dimension 3. Then, there exists a PD-decomposition

$$X \cong \overline{X} = X_0 \cup D^3, \quad X_0 \cap D^3 = S^2$$

(2.18) **Remarks** 1) As, in Propositions (2.15) one has $\pi_1(W'') \cong \pi_1(W)$ and $\pi_1(\partial W') \cong \pi_1(W')$, the pairs $(\partial Z, W')$ and (Z, W) have the same connectivity as the pairs $(X_2', \partial W')$ and (X_2, W''). In

Proposition (2.16), the pair (Z,W) has the same connectivity as the pair $(X_1,\partial W)$.

2) Proposition (2.15) was obtained by J.P.E. Hodgson [Hg], using a different method, which is valid for PD-spaces when $n \geqslant 6$. In the same paper, one finds a proof of Proposition (2.16) for PD-spaces with empty boundary.

3) Smooth 1-skeleta for PD-spaces were found by C.T.C. Wall [Wa4, Corollary 2.3.2]. The existence of a smooth 0-skeleton for a PD-spaces of formal dimension 3, in the case of empty boundary, is also established in [Wa4, Corollary 2.3.1]. Our arguments are new proofs of these results.

4) A relative version of Proposition (2.15) is given in (2.20) below. Also, the previous results work for non-finite PD-spaces, see (2.21).

For the proofs of the above proposition, we first establish, in Lemma (2.19) below, the existence of a smooth 1-skeleton for PD-spaces of formal dimension $n \geqslant 4$. Observe that this will give a proof of Proposition (2.16).

(2.19) **Lemma** Let X be a PD-space (respectively, a (Λ-PD)-space, with $Z\pi_1(X) \longrightarrow \Lambda$ locally epic) of formal dimension $n \geqslant 5$ ($n \geqslant 4$ if ∂X is empty). Then, there exists

a) a compact manifold W of dimension n, with a manifold decomposition $\partial W = W' \cup W''$ (W' may be empty). The manifold W' and the pair (W,W') admit a handle decomposition with handles of index $\leqslant 1$.

b) a PD-space (respectively, a (Λ-PD)-space) X_1 of formal dimension n with a PD-decomposition (respectively, a (Λ-PD)-decomposition) $\partial X_1 = W'' \cup X_1'$. The pair (X_1, W'') is 1-connected and so is the pair $(X_1', \partial W')$ when $W' \neq \emptyset$.

Let Z be the PD-space (respectively, a (Λ-PD)-space) $Z = X_1 \cup_{W''} W$, with boundary $\partial Z = X_1' \cup_{\partial W'} W'$.

c) a morphism of PD-spaces $F : (Z,\partial Z) \longrightarrow (X,\partial X)$ such that F and $F|\partial Z$ are homotopy equivalences (respectively, maps inducing an isomorphism on the fundamental groups and on the homology with Λ as coefficients)

Proofs of Lemma (2.19) in the case $W' = \emptyset$:

In this case, one will have $W'' = \partial W$, $X_1' = \partial X$ (whence $\partial X_1 = \partial W \amalg \partial X$) and $F|\partial X$ will be $id_{\partial X}$. In order to simplify the exposition, we give here the argument when $\partial X = \emptyset$ and indicate sometimes the slight technical changes needed for the general case. Let $X^{(n-3)}$ be the (n-3)th skeleton of X (in the general case, take $X^{(n-3)} \cup \partial X$). We assume that X is connected (otherwise, work on each component of X separately).

Let us consider the bordism group $\Omega^W_*(X,X^{(n-3)})$ of Example (1.8). By Proposition (1.7), there is a spectral sequence :

$$H_p(X,X^{(n-3)};\Omega^W_q(pt)) \Rightarrow \Omega^W_{p+q}(X,X^{(n-3)})$$

The group $\Omega^W_q(pt)$ is just the classical oriented bordism group of a point and then $\Omega^W_0(pt) = \mathbb{Z}$ and $\Omega^W_q(pt) = 0$ if $1 \leq q \leq 3$. Therefore, the term E^2 of our spectral sequence looks as follows :

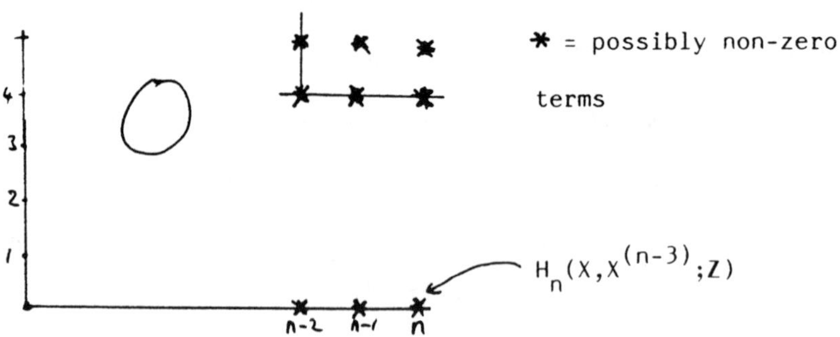

One deduces that the homomorphism $\Omega^W_n(X,X^{(n-3)}) \longrightarrow H_n(X,X^{(n-3)};\mathbb{Z})$ is an isomorphism. Therefore, there exists a manifold M^n and a map $g : (M,\partial M) \longrightarrow (X,X^{(n-3)})$ such that $g_*([M])$ is the image of $[X]$ in $H_n(X,X^{(n-3)};\mathbb{Z})$.

As X is assumed to be connected, one may assume, by adding 1-handles to ∂M or M, that M and ∂M are connected. Then perform 1 and 2-dimensional surgeries on M so that $\pi_1(M)$ $\pi_1(X)$ is an isomorphism and $\pi_2(M) \longrightarrow \pi_2(X)$ is onto. From now on, we thus assume that the pair (M,X) is 2-connected.

The handles of index ≥ 3 of M are handles of index $\leq n-3$ in the dual handle decomposition (starting from ∂M). Therefore, their image by g may be pushed by a homotopy into $X^{(n-3)}$. The restriction of g to the union of the handles of index ≤ 2 of M then represents the same element as g in $\Omega_n^w(X, X^{(n-3)})$. Thus, we shall assume that M has a handle decomposition with handles of index ≤ 2.

In the case n = 4, the diagram

$$\begin{array}{ccccccc}
\pi_2(M) & \longrightarrow & \pi_2(M,\partial M) & \longrightarrow & \pi_1(\partial M) & \longrightarrow & \pi_1(M) \\
\downarrow & & \downarrow & & \downarrow & & \downarrow \approx \\
\pi_2(X) & \longrightarrow & \pi_2(X,X^{(1)}) & \longrightarrow & \pi_1(X^{(1)}) & \longrightarrow & \pi_1(X)
\end{array}$$

shows that the homomorphism $\pi_2(M,\partial M) \longrightarrow \pi_2(X,X^{(1)})$ is surjective. When $n \geq 5$, the group $\pi_{n-2}((X,X^{(n-3)}) \cong H_{n-2}(X,X^{(n-3)};Z\pi)$ is a finitely generated $Z\pi$-module (since X has finite skeleta; we denote $\pi = \pi_1(X)$). By taking, if necessary, boundary connected sums of M with finitely many copies of $S^{n-2} \times D^2$, one can make the homomorphism

$$g_{n-2} : H_{n-2}(M,\partial M;Z\pi) \longrightarrow H_{n-2}(X,X^{(n-3)};Z\pi)$$

surjective. These two arguments permit us to assume, from now on, that the homomorphism g_{n-2} is surjective.

Form the space $Y = M \cup_{g|\partial M} X^{(n-3)}$, which is homotopy equivalent to a finite complex of dimension n. The map $g : M \to X$ and the inclusion $X^{(n-3)} \subset X$ extend to a map $f : Y \longrightarrow X$. Via the isomorphism

$$H_n(M,\partial M;Z) \xrightarrow{g_* \cong} H_n(Y,X^{(n-3)};Z) \xleftarrow{\cong} H_n(Y;Z)$$

the class [M] is mapped to a class $[Y] \in H_n(Y;Z)$ such that $f_*([Y]) = [X]$.

One has the following diagram of fundamental groups :

$$\begin{array}{ccccc}
\pi_1(\partial M) & \xrightarrow{a} & \pi_1(M) & \xrightarrow{\pi_1 g} & \\
\downarrow b & & \downarrow c & & \\
\pi_1(X^{(n-3)}) & \xrightarrow{d} & \pi_1(Y) & \xrightarrow{\pi_1 f} & \pi_1(X) = \pi
\end{array}$$

where the left hand square is a push-out diagram. In the case $n = 4$, one has by construction that $\pi_1 g$ is an isomorphism and b is onto. Therefore c and $\pi_1 f$ are isomorphisms. When $n \geq 5$, $\pi_1 g$ is an isomorphism, as well as a, d and $(\pi_1 f) \circ d$. Then, $\pi_1 f$ is an isomorphism as well as c. Thus, in all the cases $n \geq 4$, both c and $\pi_1 f$ are isomorphisms.

Let us consider the following diagram :

$$\begin{array}{ccccccc} H_{n-2}(Y;Z\pi) & \rightarrowtail & H_{n-2}(Y,X^{(n-3)};Z\pi) & \longrightarrow & H_{n-3}(X^{(n-3)};Z\pi) & \twoheadrightarrow & H_{n-3}(Y;Z\pi) \\ (*)\ \downarrow & & \downarrow & & \downarrow \simeq & & \downarrow \\ H_{n-2}(X;Z\pi) & \rightarrowtail & H_{n-2}(X,X^{(n-3)};Z\pi) & \longrightarrow & H_{n-3}(X^{(n-3)};Z\pi) & \twoheadrightarrow & H_{n-3}(X;Z\pi) \end{array}$$

where the marked surjection and isomorphisms were obtained by the choice of M. By five lemma arguments, we deduce that $H_{n-3}(Y;Z\pi) \longrightarrow H_{n-3}(X;Z\pi)$ is an isomorphism and $H_{n-2}(Y;Z\pi) \longrightarrow H_{n-2}(X;Z\pi)$ is onto. Therefore, $H_{n-1}(f;Z\pi)$ is the first non-zero homology group of the chain complex $C_*(f;Z\pi)$. This is also true when $n = 4$ since, by construction, the homomorphism $\pi_2(M) \longrightarrow \pi_2(X)$ is onto and factors through $\pi_2(Y)$.

We turn now our attention to the homology with Λ-coeficients. As the map f is (n-2)-connected, $H_{n-1}(f;\Lambda)$ is the first non-zero homology group of the chain complex $C_*(f;\Lambda)$. We claim that this is the only one. Indeed, for $i = 0,1$, one has the following diagram :

$$\begin{array}{ccccccc} H_{n-i}(M,\partial M;\Lambda) & \xrightarrow[g_*]{\simeq} & H_{n-i}(Y,X^{(n-3)};\Lambda) & \xleftarrow{\simeq} & H_{n-i}(Y;\Lambda) & \xrightarrow{f_*} & H_{n-i}(X;\Lambda) \\ \simeq \uparrow -\cap[M] & & & & \uparrow -\cap[Y] & & \simeq\uparrow -\cap[X] \\ H^i(M;\Lambda) & & \xleftarrow{\qquad\simeq\qquad} & & H^i(Y;\Lambda) & \xleftarrow{\simeq} & H^i(X;\Lambda) \end{array}$$

where the bottom arrows are isomorphisms since $\pi_1(M) \cong \pi_1(Y) \cong \pi_1(X)$. The left part of the diagram shows that

$$-\cap [Y] : H^i(Y;\Lambda) \longrightarrow H_{n-i}(Y;\Lambda)$$

is an isomorphism (for i = 0 or 1). The right part shows that

$$f_* \;:\; H_{n-i}(Y;\Lambda) \longrightarrow H_{n-i}(X;\Lambda)$$

is an isomorphism (for the same i's). Therefore, $H_*(f;\Lambda) = 0$ for $* \neq n-1$.

Similarly, for the cohomology in any Λ-module Γ, one has, for i = 0,1, the following diagram :

$$\begin{array}{ccccccc} H^{n-i}(M,\partial M;\Gamma) & \xleftarrow{\simeq} & H^{n-i}(Y,X^{(n-3)};\Gamma) & \xrightarrow{\simeq} & H^{n-i}(Y;\Gamma) & \xleftarrow{f^*} & H^{n-i}(X;\Gamma) \\ \simeq \Big\| -\cap[M] & & & & \Big\downarrow -\cap[Y] & & \simeq \Big\downarrow -\cap[X] \\ H_i(M;\Gamma) & & \xrightarrow{\qquad\qquad\simeq\qquad\qquad} & & H_i(Y;\Gamma) & \xrightarrow[\simeq]{f_*} & H_i(X;\Gamma) \end{array}$$

which shows that $f^* : H^{n-i}(X;\Gamma) \longrightarrow H^{n-i}(Y;\Gamma)$ is an isomorphism for i = 0,1. This proves that $H^n(f;\Gamma) = 0$ for any Λ-module Γ. As $H_{n-1}(f;\Lambda)$ is a finitely generated Λ-module (being the first non-zero relative homology group and Y and X having finite skeleta), the above conditions imply that $H_{n-1}(f;\Lambda)$ is a finitely generated projective Λ-module and $C_*(f;\Lambda)$ is Λ-equivalent to the (projective) chain complex

$$\cdots \longrightarrow 0 \longrightarrow H_{n-1}(f;\Lambda) \longrightarrow 0 \longrightarrow \cdots$$

As Y is a finite complex, the class of $H_{n-1}(f;\Lambda)$ in $\widetilde{K}_0(\Lambda)$ is equal to minus the Wall finiteness obstruction of $C_*(X;\Lambda)$. The latter vanishing by the definition of a (Λ-PD)-space, we deduce that $H_{n-1}(f;\Lambda)$ is stably Λ-free. One can make $H_{n-1}(f;\Lambda)$ Λ-free by attaching trivial (n-2)-cells to $X^{(n-3)} \subset Y$ and mapping them trivially into X. Still, $H_{n-1}(f;Z\pi)$ is the first non-zero homology group of f and therefore $H_{n-1}(f;\Lambda) \cong H_{n-1}(f;Z\pi)\otimes\Lambda$ (See [CS1, Lemma 1.4]). As $Z\pi_1(X)\longrightarrow\Lambda$ is locally epic, one can lift the elements of a Λ-basis of $H_{n-1}(f;\Lambda)$ into $\pi_{n-1}(f) \cong H_{n-1}(f;Z\pi)\twoheadrightarrow H_{n-1}(f;\Lambda)$ and attach (n-1)-cells to Y using the image by $\pi_{n-1}(f)\longrightarrow\pi_{n-2}(Y)$ of the elements hereby obtained. This gives a finite complex $\overline{Y}$ and f extends to $\overline{f} : \overline{Y} \longrightarrow X$ which is a Λ-equivalence inducing an isomorphism on the fundamental groups.

Let W be the union of the handles of index ≤ 1 of M. By general position, the attaching maps of the (n-1)-cells of $\overline{Y} - Y$ may pushed away from W and $\overline{Y}$ is homotopy equivalent to a space Z of the form $W \cup_{\partial W} X_1$. As $\pi_1(\partial W)\xrightarrow{\cong}\pi_1(W)$ is an isomorphism, Proposition (2.10) guarantees that $Z = W \cup X_1$ is a PD-decomposition (respectively, a (Λ-PD)-decomposition by Proposition (2.9)). Lemma (2.19) is then proven for $W' = \emptyset$. []

We shall now prove (2.15) when $W'' = \emptyset$. We use the smooth 1-skeleton already construct in Lemma (2.19). Our first argument was to construct W by finding a convenient set of 2-handles in M-intW so that the (n-1)-cells of (Y,Y) are attached away from these handles. But we give below a shorter proof due to L. Taylor.

<u>Proof of Proposition (2.15) when $W'' = \emptyset$</u> : Construct $Y = M \cup_{g|\partial M} X^{(n-3)}$ as in the proof of Lemma (2.19). One has $H_*(X,Y;\Lambda) = 0$ if $* \neq n-1$ and $H_{n-1}(X,Y;\Lambda) = H_{n-1}(X,Y;Z\pi) \otimes \Lambda$ is a finitely generated stably free Λ-module. Diagram (*) shows that

$$H_{n-1}(X,Y;Z\pi) = \ker[H_{n-2}(Y,X^{(n-3)};Z\pi) \to H_{n-2}(X,X^{(n-3)};Z\pi)],$$

and thus $H_{n-1}(X,Y;Z\pi)$ can be seen as a sub-module of $H_{n-2}(Y,X^{(n-3)};Z\pi) \cong H_{n-2}(M,\partial M;Z\pi)$.

Let M_1 be the manifold obtained from M by removing k trivial handles of index n-3 of $(U,\partial M)$ (where U is a collar neighbourhood of ∂M in M). The map $g|\partial M \cap \partial M_1$ may be extend to $g_1 : \partial M_1 \to X^{(n-3)}$ using a retraction of U onto ∂M. Replacing $(M,\partial M)$ by $(M_1 \partial M_1)$ in our construction has for unique effect to add $(Z\pi)^k$ to $H_{n-2}(M,\partial M;Z\pi)$ and to add Λ^k to $H_{n-1}(X,Y;\Lambda)$. Therefore one may suppose that $H_{n-1}(X,Y;\Lambda)$ is Λ-free.

Since the ring homomorphism $Z\pi \to \Lambda$ is locally epic, there exists elements $a_1,\dots,a_r \in H_{n-2}(X,Y;Z\pi)$ such that their images in $H_{n-2}(X,Y;\Lambda)$ constitute a Λ-basis. As seen above, the elements $a_1,\dots,a_r$ can be regarded as elements of $H_{n-2}(M,\partial M;Z\pi) = \pi_{n-2}(M,\partial M)$. Using usual handle techniques, the elements a_i may be realized by disjoint handles

$$\alpha_i : (D^{n-2}\times D^2, S^{n-3}\times D^2) \hookrightarrow (M,\partial M).$$

By construction, there are maps $\bar{\alpha}_i : D^{n-2}\times[0,1] \to X$ so that

$$\bar{\alpha}_i | D^{n-2}\times\{0\} = f \circ \alpha_i \; D^{n-2}\times\{0\}$$

$$\bar{\alpha}_i(x,t) = \alpha_i(x), \text{ for } x \in S^{n-3}\times\{0\}$$

$$\bar{\alpha}_i(D^{n-2}\times\{1\}) \subset X^{(n-3)}$$

Let $\bar{Y}$ be the complex obtained by attaching r cells of dimension n-1 to Y, using the map $\bar{\alpha}_i | \partial(D^{n-2}\times[0,1])$. The map f extends to $\bar{f} : \bar{Y} \to X$ (using the $\bar{\alpha}_i$'s) which is a homotopy equivalence (respectively, Λ-homology equivalence inducing an isomorphism on the fundamental group). Let $V = M - \text{int}(\bigcup \text{Im}\alpha_i)$. The space Y decomposes

$$Y = V \cup_{\partial V} U$$

where U retracts onto $X^{(n-3)}$. By Propositions (2.9) and (2.10), U is a PD-space (respectively, a (Λ-PD)-space) with $\partial U = \partial V$ (observe that $\pi_1(U) = \pi$ since $n \geq 5$), and the above decomposition is a PD-decomposition (respectively, a (Λ-PD)-decomposition).

Observe that M is obtained from V by adding 2-handles on ∂V. As $\pi_1(M) = \pi_1(Y)$, the homomorphism $\pi_1(V) \longrightarrow \pi_1(X)$ is surjective. On the other hand

$$H_2(X,V;\Lambda) \cong H_2(U,\partial U;\Lambda) \cong H^{n-2}(U;\Lambda) \cong H^{n-2}(X^{(n-3)};\Lambda) = 0.$$

Let (T,V,V_1) be the cobordism obtained from $V\times[0,1]$ by attaching 2 handles on $\text{int}V\times\{1\}$, using a system of generators of $\pi_2(X,V)$. Therefore, $\pi_2(X,V_1) = \pi_2(X,T) = 0$. As $H_2(X,V;\Lambda) = 0$, one has that $H_3(X,T;\Lambda) = H_3(X,V_1;\Lambda)$ is Λ-free with a basis in bijection with the handles of (T,V). One has $H_3(X,V_1;\Lambda) = H_3(X,V_1;Z\pi) \otimes \Lambda = \pi_3(X,V_1) \otimes \Lambda$, by [CS1, Lemma 1.4]. Since $Z\pi \longrightarrow \Lambda$ is locally epic, a basis of $H_3(X,V_1;\Lambda)$ can be found in the image of $\pi_3(X,V_1)$. Therefore since $n \geqslant 5$, 3-handles can be attached to T on int V_1, giving a cobordism $(T,V,\overline{V})$ such that :

- $\pi_2(X,T) = \pi_2(X,\overline{V}) = 0$
- $H_*(T,V;\Lambda) = H_*(T,\overline{V};\Lambda) = 0$

(The reader may recognize a variant of the Quillen plus construction). Let $Z = \overline{V} \cup_{\partial\overline{V}} U$. Then, Z is a PD-space and the above is a PD-decomposition (respectively, a (Λ-PD)-space...). The map f produces a map $F : Z \longrightarrow X$ which again is a homotopy equivalence (respectively, Λ-homology equivalence inducing an isomorphism on the fundamental group). In addition, the pair $(Z,\overline{V})$ is 2-connected. Let W be the union of the handles of index ≤ 2 of a handle decomposition of $\overline{V}$ and let $X_2 = Z -$

intW. Then, F, Z, W and X_2 satisfies Conditions a), b) and c) of Proposition (2.15), in the case W" = $\emptyset$. []

Proof of Lemma (2.19) and of Propositions (2.15) in the case $W' \neq \emptyset$: As $\partial(\partial X) = \emptyset$, we can apply the already established absolute case to find a smooth (1 or 2)-skeleton of ∂X. The desired results then follow from the following statement, which implies that a smooth skeleton of ∂X is always induced. []

(2.20) **Proposition** Let X be a PD-space (respectively, a (Λ-PD)-space, with $Z\pi_1(X) \longrightarrow \Lambda$ locally epic) of formal dimension $n \geqslant 5$. Let $\partial X = W' \cup R$ be a PD-decomposition (respectively, a (Λ-PD)-decomposition) of ∂X, with W' a compact manifold of dimension n-1. Then, there exists

a) a manifold W of dimension n, with a manifold decomposition $\partial W = W' \cup W''$. The pair (W,W') admits a handle decomposition with handles of index $\leqslant 2$

b) a PD-decomposition (respectively, a (Λ-PD)-decomposition)

$$X \equiv \overline{X} = W \cup X_2, \quad W \cap X_2 = W'', \quad X_2 \cap \partial X = R, \quad W \cap \partial X = W'.$$

and the pair (X,W) is 2-connected.

Proof : The inclusion $W' \subset \partial X$ represents an element α of $\Omega^W_{n-1}(\partial X, R)$. Chose a cellular decomposition of X so that R is a subcomplex. Denote by **D** the diagram :

$$\mathbf{D} = \begin{pmatrix} R & \hookrightarrow & \partial X \\ \downarrow & & \downarrow \\ R \cup X^{(n-3)} & \hookrightarrow & X \end{pmatrix}.$$

There is a commutative diagram :

$$\begin{array}{ccccc} \Omega_n^W(\mathbf{D}) \to & \Omega_{n-1}^W(\partial X,R) & \longrightarrow & \Omega_{n-1}^W(X,X^{(n-3)} \cup R) \\ & \downarrow \simeq & & \downarrow \simeq \\ & H_{n-1}(\partial X,R;\mathbf{Z}) & \longrightarrow & H_{n-1}(X,X^{(n-3)} \cup R;\mathbf{Z}) \\ & \uparrow \simeq & & \uparrow \simeq \\ & H_{n-1}(\partial X;\mathbf{Z}) & \longrightarrow & H_{n-1}(X;\mathbf{Z}) \end{array}$$

where the top line is exact. The isomorphism $\Omega_{n-1}^W(X,X^{(n-3)} \cup R) \longrightarrow H_{n-1}(X,X^{(n-3)} \cup R;\mathbf{Z})$ is guaranteed by Spectral sequence (1.7). The image of α in $H_{n-1}(\partial X,R;\mathbf{Z})$ is equal to the image of $[\partial X]$. As $[\partial X]$ goes to zero in $H_{n-1}(X;\mathbf{Z})$, one deduces that α is the image of a class in $\Omega_n^W(\mathbf{D})$. Therefore, there is a manifold M^n with a manifold decomposition $\partial M = W' \cup M'$ and a map $f : (M,W',M') \longrightarrow (X,W',X^{(n-3)} \cup R)$, with $f|W' = id_{W'}$. We can now proceed with f as in the proofs of (2.19) or (2.15), in the case $W' = \emptyset$. Everything has just to be done relative to W', replacing ∂M by M'. []

Proof of Proposition (2.17) : As in the proof of Lemma (2.19), the spectral sequence :

$$H_p(X,\partial X;\Omega^w_q(pt)) \Rightarrow \Omega^w_{p+q}(X,\partial X)$$

shows that the homomorphism $\Omega^w_3(X,\partial X) \longrightarrow H_3(X,\partial X;Z)$ is an isomorphism. Therefore, one can find a degree one map f : $(M,\partial M) \longrightarrow (X,\partial X)$, where M is a compact 3-dimensional manifold. As ∂X is a PD-space of formal dimension 2, it is homotopy equivalent to a surface [EM]. Therefore, M and f may be chosen so that $f|\partial M$ is a homotopy equivalence from ∂M to ∂X (since $\Omega^w_2(\partial X) = H_2(\partial X;Z)$). We then give the rest of the argument as if $\partial X = \emptyset$, the general case being the same.

Since f is of degree one, $\pi_1 f$ is onto and the arguments of [Wa2, Lemma 2.2] show that

a) $H_i(M;Z\pi) \longrightarrow H_i(X;Z\pi)$ is onto ($\pi = \pi_1(X)$).

b) $H_i(f;Z\pi) = 0$ if $i \neq 2,3$.

c) $H^4(f;B) = 0$ for any $Z\pi$-module B.

Add finitely many 2-cells to M, to obtain a complex M_1 and a commutative diagram :

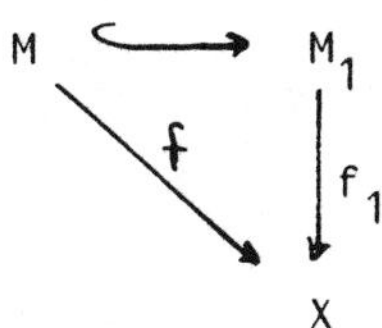

so that $\pi_1 f_1$ is an isomorphism. One checks that the map f_1 satisfies Conditions a) b) and c) above with in addition $H_2(f_1;Z\pi) = 0$. It follows from [Wa2, Lemma 2.3] that $H_3(f_1;Z\pi) = \pi_3(f_1)$ is projective. As M_1 and X are finite, $\pi_3(f_1)$ is stably free and may be made free by adding to M_1 trivial 2-cells sent to the base point of X. We thus get a factorization

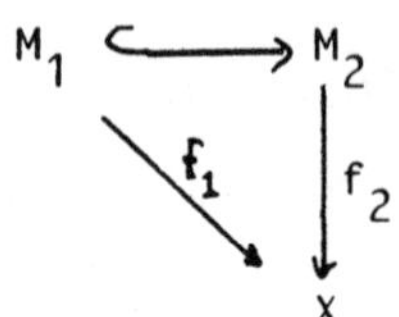

By adding 3-cells to M_2 according to a free basis of $\pi_3(f_2)$, one obtain a homotopy equivalence $M_3 \xrightarrow{\simeq} X$. By general position, the attaching maps of these 3-cells can avoid a 3-disk D of $M \subset M_2$. Set $X = M_3$ and $X_1 = X - \text{int}D$. By (2.10) the decomposition $X = X_1 \cup D$ is a PD-decomposition. []

The fact that a PD-space is homotopy equivalent to a finite complex does not play an essential role in the proofs of Propositions (2.15), (2.16) and (2.18). One can generalize them to PD-spaces which are just dominated by a finite complex, as done in (2.21) below. For applications, see Section 8.C.

(2.21) **Proposition** Statements (2.15) to (2.20) are true in the category of finitely dominated PD-spaces.

Proof : The proofs of (2.19), (2.20) and (2.17) will be the same as for the original version, till the use of the Wall finiteness obstruction, to kill $H_{n-1}(f;\Lambda)$ (or $H_3(f_1;Z\pi)$ in the proof of (2.17)). This last homology group is only projective and not stably free. But it can be made (infinitely generated) free by adding to Y cells of dimension n-2 and n-1 (or 2 and 3 for (2.17)) without touching the manifold part of the source complex of f (or f_1). Therefore, f (or f_1) can be made a homotopy equivalence by adding (infinitely many) (n-1)-cells (or 3-cells) to the source complex. The last argument of general position in the proofs of (2.19) and (2.17) works as well.

For Proposition (2.15), since $n \geqslant 5$, one has $\pi_1(X^{(n-3)}) = \pi_1(X)$. Therefore one may choose a representative of the homotopy type of X such that $(X, X^{(n-3)})$ is homotopy equivalent to a finite pair (the Wall finiteness obstruction of X being already realized in $X^{(n-3)}$). The proof then goes through as in the finite case. []

3. Q-SPACES - POINCARE SPACES

In this section, some formal material, used in future chapters, is developed. It generalizes an idea of Quinn [Qu] : the expression of the homology groups, $H_n(U,V;MG)$, as bordism groups of objects (Q-spaces) which enjoy some properties of manifolds.

A. A HOMOLOGY THEORY OVER BG

Let BG_k be the classifying space for (k-1)-spherical fibrations and let BG be the inductive limit of the BG_k's (see [MM, Chapter 3]). The universal (k-1)-spherial fibration over BG_k is denoted by $E_k \longrightarrow BG_k$. If $j : B \longrightarrow BG_k$ is a map, the following notations will be used for the induced fibration over B :

- $E(j)$ for its total space
- $\hat{E}(j)$ for the mapping cylinder of the projection $E(j) \longrightarrow B$
- $T(j) = \hat{E}(j)/E(j)$, its Thom space.

Our working category is the category of **pairs over** BG. An object of this category is a pair (X,Y) of a CW-complex X and a subcomplex Y of X together with a map $j^X : X \longrightarrow BG$ (usually not explicitely mentioned in the notation). A morphism from

(X,Y) to (X',Y') is a map $f : X \longrightarrow X'$ such that $f(Y) \subset Y'$ and $j^{X'} \circ f = j^X$ (equality as maps). One denotes by j^Y the restriction of j^X to Y. The pair $(X,\emptyset)$ over BG is treated as a single space and we shall say : **a space** X **over** BG.

It is classical that $\pi_1(BG) = \{\pm 1\}$. Let us chose once a map $w^{BG} : BG \longrightarrow \mathbf{R}P^\infty$ inducing an isomorphism on the fundamental groups. The every pair over BG becomes a pair over $\mathbf{R}P^\infty$.

Let (X,Y) be a pair over BG. Define :

$$X_k = (j^X)^{-1}(BG_k) \subset X \ , \quad j^X_k = j^X\big|X_k$$

$$Y_k = X_k \cap Y = (j^Y)^{-1}(BG_k) \ , \ j^Y_k = j^Y\big|Y_k$$

The collection of Thom spaces $\{T(j^X_k)\}_{k \in \mathbf{N}}$ constitutes a spectrum $T(j^X)$ which is called the **Thom spectrum** of j^X. For instance, if $X = BG$ and $j^X = id$, the spectrum $T(j^X)$ is the clasical Thom spectrum MG. We are interested in the groups :

$$\mathcal{H}_n(X,Y) := \pi_n(T(j^X),T(j^Y)) = \lim_k \pi_{n+k}(T(j^X_k),T(j^Y_k))$$

As stable homotopy is a homology theory, one checks that $\mathcal{H}_*(X,Y)$ is a homology theory over BG, in the sense of [Do, Section 2].

(3.1) <u>**Examples**</u> : a) Let U be a CW-complex and V be a subcomplex of U. Let $(X,Y) = (U\times BG, V\times BG)$, made a pair over BG via the natural projection $j^X : U\times BG \rightarrow BG$. The restriction

$j_k^X : U{\times}BG_k \longrightarrow BG_k$ is also the natural projection. Thus, one has

$$E(j_k^X) = U{\times}E_k, \quad E(j_k^X) = U{\times}\hat{E}_k,$$
$$T(j_k^X)/T(j_k^X|V{\times}BG_k) = U{\times}\hat{E}_k/U{\times}E_k \cup \hat{V}{\times}E_k = U/(V\wedge\hat{E}_k/E_k).$$

Therefore

$$\mathcal{H}_n(X,Y) = \lim_k \pi_{n+k}(U/(V\wedge\hat{E}_k/E_k)) = H_n(U,V;MG).$$

b) Applying the same argument, one obtains

X	Y	$\mathcal{H}_n(X,Y)$
$U{\times}BSG$	$V{\times}BSG$	$H_n(U,V;MSG)$
$U{\times}BSO$	$V{\times}BSO$	$H_n(U,V;MSO)$
$U{\times}BO$	$V{\times}BO$	$H_n(U,V;MO)$
$U{\times}pt$	$V{\times}pt$	$\pi_n^S(U,V)$

Here, j^X is the composition of the natural maps $U{\times}BSO \longrightarrow U{\times}BG$, $U{\times}BO \longrightarrow U{\times}BG$, etc, with the projection $U{\times}BG \longrightarrow BG$.

(3.2) Let (X,Y) be a pair over BG. Consider the classical Hurewicz homomorphism :

$$\pi_{n+k}(\hat{E}(j_k^X),E(j_k^X)\cup\hat{E}(j_k^Y)) \longrightarrow \dot{H}_{n+k}(\hat{E}(j_k^X),E(j_k^X)\cup\hat{E}(j_k^Y);Z)$$

(recall that $\dot{H}_*$ denotes the standard singular homology with local coefficients while our working homology over $\mathbf{R}P^\infty$ is denoted by H_*).

On the other hand, the standard orientation of (D^k,S^{k-1}) gives a Thom class $u_k^s \in \dot{H}^k(E_k^s,E_k^s;Z)$, where $E_k^s = E(j_k^{BSG})$, the natural k-spherical fibration over BSG_k. Using the diagram :

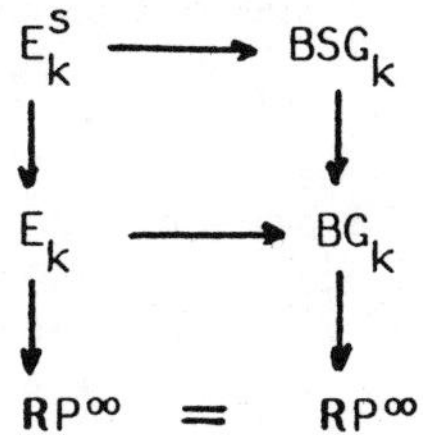

and the Shapiro Lemma, one obtains :

$$\dot{H}^k(\hat{E}_k^s,E_k^s;Z) = \dot{H}^k(\hat{E}_k,E_k;\mathrm{Hom}_Z(Z\{\pm 1\};Z))$$

where the $Z\{\pm 1\}$-module structure on $\mathrm{Hom}_Z(Z\{\pm 1\};Z)$ is given by $(g\alpha)(b) = \alpha(gb)$, $g \in \{\pm 1\}$ and $\alpha \in \mathrm{Hom}_Z(Z\{\pm 1\};Z)$. Thus, $\mathrm{Hom}_Z(Z\{\pm 1\};Z) = Z^w$, where the latter denotes the abelian group Z with the $Z\{\pm 1\}$-module structure given by $(-1)z = -z$. Denote by u_k the image of u_k^s in $H^k(\hat{E}_k,E_k;Z^w)$. The cap-product with $(j_k{}^X)^*(u_k)$ gives the Thom isomorphism

$$t_k : \dot{H}_{n+k}(\hat{E}(j_k^X),E(j_k^X)\cup\hat{E}(j_k^Y);Z) \longrightarrow$$

$$\longrightarrow \dot{H}_n(\hat{E}(j_k^X),\hat{E}(j_k^Y);Z^w \otimes Z) = H_n(X,Y;Z).$$

Composing t_k with h and passing to the inductive limit on k gives the homomorphism :

$$t : \mathcal{H}_n(X,Y) \longrightarrow H_n(X,Y;Z) .$$

B. Q-SPACES

We now define the objects used to express $\mathcal{H}_n(X,Y)$ in terms of a bordism group. By definition, a **Q-space** (of **formal dimension** n) consists of :

1) a CW-complex Q over BG with a prescribed subcomplex ∂Q.
2) a class $(Q) \in \mathcal{H}_n(Q,\partial Q)$. (the $\mathcal{H}$-fundamental class)

We abbreviate by Q the data $(Q,\partial Q,j^Q,(Q))$. The **boundary** of Q is the Q-space ∂Q of formal dimension n-1, where it is understood that $\partial(\partial Q) = \emptyset$ and (∂Q) is the image of (Q) under the boundary homomorphism $\mathcal{H}_n(Q,\partial Q) \longrightarrow \mathcal{H}_{n-1}(\partial Q)$.

Q-spaces were introduced by F.Quinn [Qu] and were called "normal spaces". Because of our extensive use of standard surgery theory, we found this terminology slightly confusing. We thus propose the shorter name of "Q-spaces", taking the initial of their inventor.

(3.2) <u>**Example**</u> Let M^n be a compact smooth manifold with boundary ∂M. If $k > n$, there is an embedding $(M,\partial M) \hookrightarrow (D^{n+k},S^{n+k-1})$ which is unique up to isotopy, with normal bundle ν_k. Consider $(M,\partial M)$ as a pair over BG by choosing a characteristic map $j_k^M : M \longrightarrow BG_k$ of the sphere bundle associated to ν_k. The Thom-Pontrjagin class of $(M,\partial M)$ in $\pi_{n+k}(T(j_k^M);T(j_k^{\partial M}))$ provides a stable class $(M) \in \mathcal{H}_n(M,\partial M)$. This will be the natural way to see a manifold as a Q-space.

As in the case of PD-spaces, two simple operations on Q-spaces (of type "cutting and pasting") are now described. In this context, they are actually easier.

(3.3) Cutting Q-spaces : Let Q be a Q-space of formal dimension n. Let us consider a CW-decomposition :

$$Q = Q_1 \cup Q_2 , \quad Q_1 \cap Q_2 = Q_0 , \quad R_i = Q_i \cap \partial Q \quad (i = 0,1,2)$$

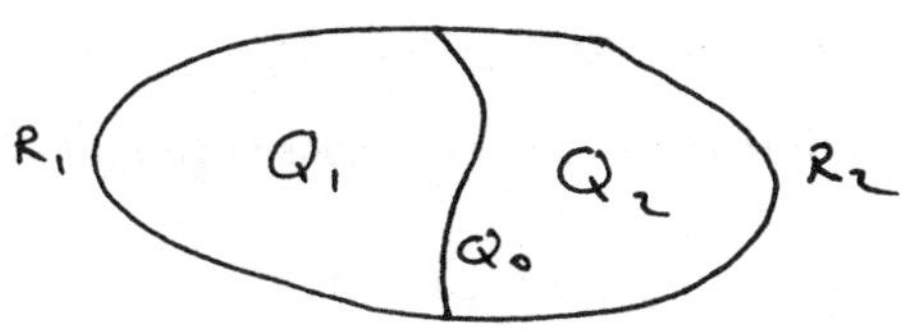

This determines a structure of a Q-space of formal dimension n on Q_1 and Q_2 by setting $\partial Q_i = R_i \cup Q_0$ (i = 1,2) and $(Q_i) = s_i((Q))$, where s_i are the homomorphisms :

$$s_1 : \mathcal{H}_n(Q,\partial Q) \longrightarrow \mathcal{H}_n(Q,\partial Q \cup Q_2) \cong \mathcal{H}_n(Q_1,\partial Q_1)$$

$$s_2 : \mathcal{H}_n(Q,\partial Q) \longrightarrow \mathcal{H}_n(Q,\partial Q \cup Q_1) \cong \mathcal{H}_n(Q_2,\partial Q_2).$$

On the other hand, one obtains three Q-spaces of formal dimension n-1 (R_1, R_2 and Q_0) by setting $\partial R_1 = \partial R_2 =$ $= \partial Q_0 = Q_0 \cap \partial Q$ and :

(R_1) = image of (∂Q) under the homomorphism

$$\mathcal{H}_{n-1}(\partial Q) \to \mathcal{H}_{n-1}(\partial Q, R_2) \cong \mathcal{H}_{n-1}(R_1, \partial R_1)$$

(R_2) = image of (∂Q) under the homomorphism

$$\mathcal{H}_{n-1}(\partial Q) \to \mathcal{H}_{n-1}(\partial Q, R_1) = \mathcal{H}_{n-1}(R_2, \partial R_2)$$

(Q_0) = image of (∂Q_1) under the homomorphism

$$\mathcal{H}_{n-1}(\partial Q_1) \to \mathcal{H}_{n-1}(\partial Q_1, R_1) \cong \mathcal{H}_{n-1}(Q_0, \partial Q_0)$$

Observe that exchanging the role of Q_1 and Q_2 changes the sign of (Q_0). We say that the spaces Q_1, Q_2, etc, with their Q-space structures constitute a **Q-decomposition** of Q.

Pasting Q-spaces is slightly less canonical :

(3.4) **Lemma** Let $(Q,\partial Q)$ be a pair over BG with a CW-decomposition as in (3.3). Consider the homomorphisms :

$$r_i : \mathcal{H}_n(Q_i, \partial Q_i) \to \mathcal{H}_{n-1}(\partial Q_i) \to \mathcal{H}_{n-1}(\partial Q_i, R_i) \cong \mathcal{H}_{n-1}(Q_0, \partial Q_0)$$

s_i : as in (3.3)

Let $(Q_i) \in \mathcal{H}_n(Q_i, \partial Q_i)$ such that $r_i((Q_1)) = -r_2((Q_2))$. Then there is a class $(Q) \in \mathcal{H}_n(Q,\partial Q)$ such that $s_i((Q)) = (Q_i)$. If $(Q)' \in \mathcal{H}_n(Q,\partial Q)$ also satisfies $s_i((Q)') = (Q_i)$, then $(Q)' - (Q) \in \mathrm{Im}[\mathcal{H}_n(Q_0, \partial Q_0) \to \mathcal{H}_n(Q,\partial Q)]$.

Proof : The proof is the same as the beginning of the proof of Proposition (2.5) []

(3.5) A **Q-cobordism** between two Q-spaces Q_1 and Q_2 consists of :

1) a Q-space A of formal dimension n+1

2) a CW-decomposition $\partial A = Q_1 \cup A_0 \cup Q_2$, $Q_i \cap A_0 = \partial Q_i$, $Q_1 \cap Q_2 = \emptyset$, such that $\mu((A)) = ((Q_1),-(Q_2))$, where μ is the following homomorphism :

$$\mu : \mathcal{H}_{n+1}(A;\partial A) \longrightarrow \mathcal{H}_n(\partial A) \longrightarrow \mathcal{H}_n(\partial A, A_0)$$
$$\downarrow \simeq$$
$$\mathcal{H}_n(Q_1,\partial Q_1) \oplus \mathcal{H}_n(Q_2,\partial Q_2)$$

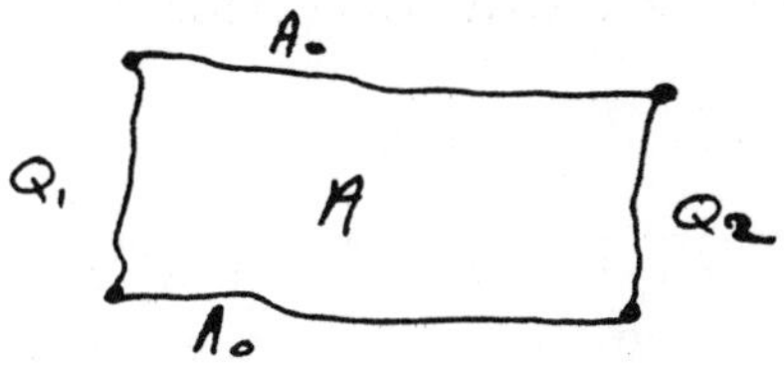

A_0

Q_1 A Q_2

A_0

Using Lemma (3.4), one checks that the Q-bordism is an equivalence relation.

(3.6) Let (X,Y) be a pair over BG. The **Q-bordism group** $\Omega_n^Q(X,Y)$ **of** (X,Y) is the group of equivalence classes of pairs (Q,f), where :

1) Q is a Q-space of formal dimension n

1) $f : (Q,\partial Q) \longrightarrow (X,Y)$ is a map over BG (i.e. $j^Q = j^X \circ f$).

Two such pairs (Q_1,f_1) and (Q_2,f_2) are considered as equivalent if there is a Q-cobordism A between them and a map $F : (A,A_0) \longrightarrow (X,Y)$ such that $F|Q_i = f_i$. The group law on $\Omega_n^Q(X,Y)$ is given by the disjoint union.

We shall now establish several properties of the groups $\Omega_n^Q(X,Y)$. Let f_0, $f_1 : (Q,\partial Q) \longrightarrow (X,Y)$ be maps and let $F : (Q\times[0,1],\partial Q\times[0,1]) \longrightarrow (X,Y)$ be a homotopy between them. The lifting properties for homotopies gives a homotopy equivalence $TF : T(j^X\circ f_0)\times[0,1] \longrightarrow T(j^X F)$ extending the inclusion $T(j^X\circ f_0)\times 0 \subset T(j^X\circ F)$. Define $T_1F : T(j^X\circ f_0) \longrightarrow T(j^X\circ f_1)$ as $TF_1(u) = TF(u,1)$. Let Q' be the Q-space obtained by changing the $\mathcal{H}$-fundamental class of Q into $(Q') = T_1F((Q))$.

(3.7) **Lemma** The pairs (Q,f_0) and (Q',f_1) represent the same element in $\Omega_n^Q(X,Y)$.

Proof : The required Q-cobordism (A,F) may be chosen as follows : $A = Q\times[0,1]$, $A_0 = \partial Q\times[0,1]$, F is the considered homotopy and (A) is represented by :

$$\beta : (D^{n+k}\times[0,1],S^{n+k-1}\times[0,1]) \longrightarrow (T(j_k^X\circ F),T(j_k^X\circ F|\partial A))$$

defined as $\beta(x,t) = TF(\alpha(x),t)$. []

(3.8) **Lemma** Let (Q,f) represent an element in $\Omega_n^Q(X,Y)$. Let us consider a pair $(Q_1,\partial Q_1)$ and a commutative diagram :

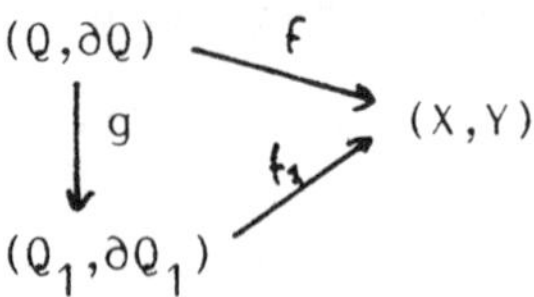

Make Q_1 a Q-space by setting $(Q_1) = Tg((Q))$. Then, the pairs (Q,f) and (Q_1,f_1) represent the same element in $\Omega_n^Q(X,Y)$.

Proof : Set $A = \mathcal{M}_g$, the mapping cylinder of g, and $A_0 = \mathcal{M}_{g|\partial Q}$. Define $F : A \longrightarrow X$ by $F(q,t) = f(q)$ for $(q,t) \in Q\times[0,1]$ and $F|Q_1 = f_1$.

The inclusion $Q\times[0,1] \subset A$ extends to an inclusion $T(j_k^X f)\times[0,1] \subset T(j_k^X F)$. The $\mathcal{H}$-fundamental class (A) shall be represented by the composed map :

$$D^{n+k}\times[0,1] \xrightarrow{(Q)\times id} T(j_k^X f)\times[0,1] \longrightarrow T(j_k^X F). \qquad []$$

The Q-bordism group is just the construction which enables us to represent formally the homology theory $\mathcal{H}_n$ under the form of a bordism group. More precisely :

(3.9) **Proposition** Let (X,Y) be a pair over BG. For all n, the map

$$\eta : \Omega_n^Q(X,Y) \longrightarrow \mathcal{H}_n(X,Y)$$

given by $(Q,f) \longmapsto Tf((Q))$ is an isomorphism.

Proof : We must first check that η is well defined. Let (A,F) be a Q-cobordism between (Q_1,f_1) and (Q_2,f_2). Let us consider the following commutative diagram :

$$\begin{array}{ccccc} \mathcal{H}_{n+1}(A,\partial A) & \longrightarrow & \mathcal{H}_n(\partial A) & \longrightarrow & \mathcal{H}_n(A) \\ & & \downarrow & & \downarrow \\ & & \mathcal{H}_n(\partial A, A_0) & & \mathcal{H}_n(A,\partial A) \end{array}$$

The difference $(Q_2) - (Q_1)$ viewed as an element of $\mathcal{H}_n(\partial A, A_0)$ has image zero in $\mathcal{H}_n(A,\partial A)$, since the top line is exact and $(Q_2) - (Q_1) = \mu((A))$ by the definition of (A) in (3.5).

Therefore, (Q_1) and (Q_2) have same image in $\mathcal{H}_n(X,Y)$ and η is well defined.

Let $\gamma \in \mathcal{H}_n(X,Y)$. Then $\gamma = \eta(X,id_X)$ (with $\partial X = Y$), which shows that η is surjective. By Lemma (3.8), the pairs (Q,f) and (X,id_X), with $(X) = Tf((Q))$, represent the same element of $\Omega_n^Q(X,Y)$; this shows the injectivity of η. []

(3.10) **Examples** Let U be a CW-complex and V be a subcomplex of U. Let us consider the pair $(U\times BG, V\times BG)$ as a pair over BG via the natural projection $j^{U\times BG} : U\times BG \longrightarrow BG$. Using (3.1) and (3.9), one obtains

$$\Omega_n^Q(U\times BG, V\times BG) \cong H_n(U,V;MG).$$

We leave to the reader to complete the list of examples as in (3.1). The group $\Omega_n^Q(U\times BG, V\times BG)$ is the group $\Omega_n^{norm}(U,V)$ (bordism group of "normal spaces") which was denoted by $\Omega_n^N(U,V)$ in [Qu], where it is defined as follows : the elements of $\Omega_n^{norm}(U,V)$ are bordism classes of 4-tuples (Q,f,μ,α) where

1) Q is a CW-complex with a subcomplex ∂Q

2) $f : (Q,\partial Q) \longrightarrow (U,V)$ is a map

3) $\mu : Q \longrightarrow BSG$ is a map

4) $\alpha \in \pi_n(T(\mu),T(\mu\ V))$.

The isomorphism $\Omega_n^{norm}(U,V) \longrightarrow \Omega_n^Q(U\times BG, V\times BG)$ is given by

$$(Q,f,\mu,\alpha) \longmapsto \{(Q,(f,\mu)),\ j^Q = \mu,\ (Q) \text{ represented by } \alpha\}$$

Thus, our bordism groups $\Omega_n^Q(X,Y)$ are an extension to a more general setting of the normal bordism groups of [Qu]. The advantage of this gain in generality will appear in later chapters, where we shall play with various liftings of j^X, as in the following example : let (X,Y) be a pair over BG and let $\tilde{j}^X : X \longrightarrow BO$ be a lifting of j^X. Let $\Omega_n^M(X,Y;\tilde{j}^X)$ be the group of bordism classes of pairs (M,f), where

1) M is a compact smooth manifold of dimension n

2) $f : (M,\partial M) \longrightarrow (X,Y)$ is a continuous map

3) $\tilde{j}^X \circ f$ classifies the normal bundle of M.

(3.11) **Proposition** For any lifting $\tilde{j}^X$ of j^X as above, the correspondence :

$$(M,f) \longmapsto \{(M,f),\ j^M = \tilde{j}^X \circ f,\ (M) = \text{Thom-Pontrjagin class of } f\}$$

provides an isomorphism from $\Omega_n^M(X,Y;\tilde{j}^X)$ onto $\Omega_n^Q(X,Y)$.

Proof : By Proposition (3.9) and its proof, each element of $\Omega_n^Q(X,Y)$ admits a representative of the form $(X,id,\bar{\alpha})$. As $\mathcal{H}_*$ is a homology theory, α is in the image of $\alpha \in \mathcal{H}_n(X^{(m)},Y^{(m)})$, for m large enough, where $X^{(m)}$ is the m-skeleton of X. By (3.9) again, each element of $\Omega_n^Q(X,Y)$ admits a representative of the form $(X^{(m)},\text{inclusion},\alpha)$. The diagram

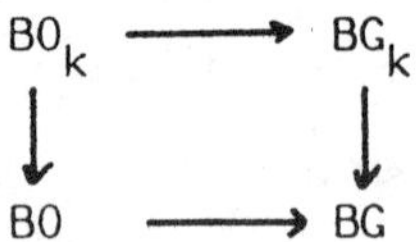

is (2k-2)-connected (see [Hr]). For k large enough, the lifting $\hat{j}^X$ induces a lifting $\tilde{j}_k : (X^{(m)})_k \longrightarrow BO_k$. This gives us a k-vector bundle ξ_k over $(X^{(m)})_k$ and $T(\xi_k) = T(\tilde{j}_k)$. If k is large enough, the class (X) is representable by a map $\alpha : (D^{n+k},S^{n+k-1}) \longrightarrow (T(\xi_k),T(\xi_k|(Y^{(m)})_k)$. Making α transverse to $(X^{(m)})_k$ (considered as the zero section of ξ_k) produces a manifold $M = \alpha^{-1}((X^{(m)})_k)$. Set $g = \alpha|M$. One has $\nu_M = g^*(\xi_k)$ and $(X) = Tg((M))$, where (M) is the Thom-Pontrjagin class of M. By Lemma (3.8), (M,g) and (X,id_X) represent the same element of $\Omega_n^Q(X,Y)$, which proves the surjectivity of our correspondence. A similar use of transversality gives the injectivity. □.

(3.12) Another consequence of Lemma (3.8) is what is called "surgery in Q-spaces". Let (Q,f) represent an element of $\Omega_n^Q(X,Y)$. Let $t \in \pi_{i+1}(f)$ be represented by a commutative diagram :

$$\begin{array}{ccc} S^i & \xrightarrow{\tau} & Q \\ \cap & & \downarrow \\ D^{i+1} & \xrightarrow{\overline{\tau}} & X \end{array}$$

Form the space $\overline{Q} = Q \cup_{\tau}$(cell of dim. i+1). Using $\overline{\tau}$ on the additional cell, the map f extends to $\overline{f} : Q \longrightarrow X$. Denote by g the inclusion $Q \subset \overline{Q}$. By Lemma (3.8), the pairs (Q,f) and $(\overline{Q},\overline{f})$ represent the same element of $\Omega_n^Q(X,Y)$ (setting $(\overline{Q}) = Tg((Q))$ and $\overline{\partial Q} = \partial Q$). Observe that, by (3.8), the mapping cylinder of g constitutes a Q-cobordism between Q and $\overline{Q}$.In further chapters, this operation will be used in connection with surgery in Poincaré spaces and play the same role (but is much more elementary). For this reason, we say that $(\overline{Q},\overline{f})$ is obtained from (Q,f) by **surgery (along** t or τ). This operation can also be performed with a class $t \in \pi_{i+1}(f|\partial Q)$ (surgery on ∂Q).

We shall end this subsection with the following result :

(3.13) **Proposition** Let (X,Y) be a pair over BG. Then there is a spectral sequence :

$$E_2^{p,q} = H_p(X,Y;\pi_q^s) \Longrightarrow \Omega_n^Q(X,Y)$$

where the stable homotopy groups of spheres π_q^s are considered as trivial $\pi_1(X)$-modules.

<u>Proof</u> : By (3.9), $\Omega_n^Q(X,Y)$ can be replaced by $\mathcal{H}_n(X,Y)$, and $\mathcal{H}_q(pt) = \pi_q^s$ as an abelian group. Then, by (2.5), it suffices to see that $\mathcal{H}_*$ is a w-homology theory.

Let $\gamma : [0,1] \longrightarrow X$ be a loop at $a \in X$ and let $\alpha : S^{n+k} \longrightarrow S^k = T(j^{\{a\}})$ represent an element $[\alpha]$ of $\mathcal{H}_n(\{a\})$. According to [Do, 2.4], the class $\beta[\alpha] \in \mathcal{H}_n(\{a\})$ is represented by $\beta| S^{n+k}\times 1$, where :

$$\beta : S^{n+k}\times[0,1] \xrightarrow{\alpha\times id} S^k\times[0,1] = T(\gamma\circ j^X) \xrightarrow{T\gamma} T(j^X)$$

But the degree of the two maps :

$$T(\gamma|0) : S^k\times 0 \longrightarrow T(j^{\{a\}}) = S^k$$

$$T(\gamma|1) : S^k\times 1 \longrightarrow T(j^{\{a\}}) = S^k$$

differ by the multiplication by $w^X(\gamma)$. Therefore $\gamma[\alpha] = w^X(\gamma)[\alpha]$ and $\mathcal{H}_*$ is a w-theory. []

C. POINCARÉ SPACES

A Q-space P of formal dimension n is called a **Poincaré Space** (of formal dimension n) if P is a PD-space with $t((P)) = [P]$, where $t : \mathcal{H}_n(P,\partial P) \longrightarrow H_n(P,\partial P;Z)$ is the homomorphism of (3.2). The notion of **Poincaré cobordism** is defined accordingly, as in (3.5). (the more general notion of Λ-Poincaré space will be studied in Section 8.B).

Observe that j^P is then a classifying map for the Spivak bundle of P and therefore we call the pair $(j^P,(P))$ a **Spivak pair** for P. The existence of a Spivak pair and its unicity up to fibre equivalence is classical (see [Sp] or [Br1]) and implies that every PD-space admits the structure of a Poincaré space which is unique up to Poincaré cobordism.

A **morphism of Poincaré spaces** between two Poincaré spaces R and P is a map $f : R \longrightarrow P$ over BG which is a **morphism of Q-spaces**, that is $\mathcal{H}_n f((R)) = (P)$. This implies that $H_*(f;Z)([R]) = [P]$ and thus f is a morphism of PD-spaces. Morphisms of Poincaré spaces are the generalization to Poincaré spaces of the concept of "normal map of degree one" which is used in classical surgery theory.

A **Poincaré decomposition** of a Poincaré space P is a Q-decomposition :

$P = P_- \cup P_+$, $\partial P = \partial_- P \cup \partial_+ P$,

$P_- \cap P_+ = P_0$, $\partial P_\pm = \partial_\pm P \cap P_0$

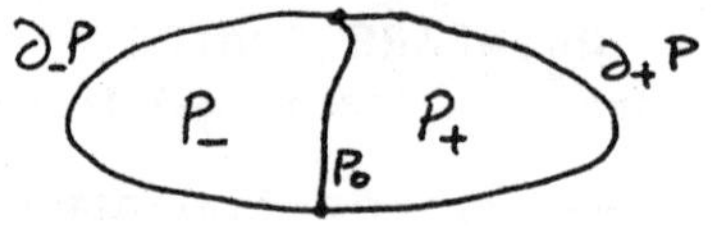

such that every space under consideration is a Poincaré space.

Let Z be a Poincaré space of formal dimension k and P be a Poincaré space of formal dimension n. A **Poincaré embedding** (of codimension n-k) of Z into P consists of :

a) a $(n\text{-}k\text{-}1)$-sphere bundle ν over Z, classified by $\nu : Z \longrightarrow BG_{n-k}$. Let $(\hat{E},E) = (\hat{E}(\nu),E(\nu))$.

b) a map $j^{\hat{E}} : \hat{E} \longrightarrow BG$ characteristic for $j^X + \nu$ (we use the H-space structure on BG given by the join operation (see [MM,1.19])). Consider the isomorphism

$$\mu : \mathcal{H}_n(P,\partial P) \longrightarrow \mathcal{H}_n(Z,\partial Z) \text{ given by :}$$

$$\begin{aligned}\mu : \mathcal{H}_n(P,\partial P) &= \varinjlim \pi_{n-r}(\hat{E}(j_r^{\hat{E}}),E(j_r^{\hat{E}}) \cup \hat{E}(j_r^{\partial\hat{E}}) = \\ &= \varinjlim \pi_{n+r}(\hat{E}(j_{r+n-k}^{Z}),E(j_{r+n-k}^{Z}) \cup \hat{E}(j_{r+n-k}^{\partial Z})) = \\ &= \mathcal{H}_k(Z,\partial Z).\end{aligned}$$

Consider E as a Poincaré space by setting $(E) = \mu^{-1}((Z))$.

c) a Poincaré space $\overline{P}$ of formal dimension n and a Poincaré decomposition $\overline{P} = \overline{P}_- \cup \overline{P}_+$, etc, of P, with $(\hat{E},\hat{E}(\nu|\partial Z),E) = (\overline{P}_-,\partial_-\overline{P},\overline{P}_0)$ as triple of Poincaré spaces.

d) a morphism of Poincaré spaces $\alpha : \overline{P} \longrightarrow P$ which is a homotopy equivalence.

The cases $\partial Z = \emptyset$ and $\partial Z = \partial P = \emptyset$ are of course possible. The bundle ν is the **normal bundle** of the Poincaré embedding. If $f : Z \longrightarrow P$ is a map over BG and if there is a Poincaré embedding as above such that $\alpha|Z = f$, we then say that f is equivalent to (or can be realized by) a Poincaré embedding.

(3.14) **Lemma** Let Z be a Poincaré space of formal dimension k and P be a Poincaré space of formal dimension n, with $n \geqslant k+3$. Let $f : Z \longrightarrow P$ be a map over BG. If f can be realized by a PD-embedding, then f can be realized by a Poincaré embedding.

Proof : We give here the particular case where $\partial P = \emptyset$, the general case being similar. Since f can be realized by a PD-embedding, one can write $P = \hat{E} \cup P_+$, $\hat{E} \cap P_+ = E$, with $\nu : E \longrightarrow Z$ being a (n-k-1)-spherical fibration. It is possible to suppose that E and P_+ are finite complexes. Let :

- $(W_{\hat{E}}, W_E)$ be regular neighbourhoods of $(\hat{E}, E)$ embedded into (D^{n+s}, S^{n+s-1})
- (W_+, W_E) be regular neighbourhoods of (P_+, E) embedded into (D^{n+s}, S^{n+s-1})

Then, $W = W_{\hat{E}} \cup W_+$ (union over W_E) is a regular neighbourhood of P embedded into S^{n+s}. The map $\partial_+ W_{\hat{E}} = W_{\hat{E}} \cap \partial W \longrightarrow E$ (the restriction of the collapse $W_{\hat{E}} \searrow \hat{E}$), considered as a Serre fibration, has fiber S^{s-1} and represents the Spivak bundle of $\hat{E}$ (see [Br2]); $W_{\hat{E}}$ may be considered as the disk bundle associated to $\partial W_{\hat{E}} \longrightarrow \hat{E}$. The same holds true for $\partial N \longrightarrow Z$, where

N is a regular neighbourhood of Z embedded into S^{k+s}. As f is over BG, it extends to a map $F : (N,\partial N) \longrightarrow ((W_{\hat{E}},\partial_+ W_{\hat{E}})$, such that F and $F|\partial N$ are simple homotopy equivalences. As $n \geq k+3$, one may use [Wa2, Corollary 11.3.1 and 11.3.1 relative], to homotop F to a PL-embedding, so that $W_{\hat{E}}$ is a regular neighbourhood of F(N).

We have now a new PD-embedding of Z into P, given by :

$\overline{P} = W$, Poincaré decomposed in $W = W_{\hat{E}} \cup W_+$,

α = collapse of W onto P,

$\nu : W_E \subset W_{\hat{E}} \searrow N \searrow Z$ being a (n-k-1)-spherical fibration.

Now, $W_{\hat{E}}$ may be considered as a regular neighbourhood of Z embedded into S^{n+s}. As the collapsing maps

$$S^{n+s} \longrightarrow S^{n+s}/(S^{n+s} - \mathrm{int}W) = W/\partial W$$

$$S^{n+s} \longrightarrow S^{n+s}/(S^{n+s} - \mathrm{int}W_{\hat{E}}) = W_{\hat{E}}/\partial W_{\hat{E}}$$

represent $(\overline{P})$ and (Z), the relationship between these two classes guarantees that the new PD-embedding is a Poincaré embedding. $\square$

D. POINCARE SURGERIES : PRELIMINARIES

Let X be a space over BG. Let $f : P \longrightarrow X$ be a map over BG, where P be Poincaré space of formal dimension n. We shall often use, for the group $\pi_j(f)$, the notation $\widetilde{K}_j(f)$ (homotopy kernels) or even $\widetilde{K}_j(P)$ when there is no ambiguity. An element $b \in \widetilde{K}_j(f)$ is represented by a commutative diagram

$$\begin{array}{ccc} S^j & \xrightarrow{\beta} & P \\ \cap & & \downarrow f \\ D^{j+1} & \xrightarrow{\beta} & X \end{array}$$

We denote by b the class of β in $\pi_j(P)$. Suppose that b can be realized by a Poincaré embedding with trivialized normal bundle. This means that there exists

i) a Poincaré embedding $(\nu,\hat{E},\hat{P},\alpha,\text{etc})$ as described in (2.11) of S^j into P (with the $\mathcal{H}$-fundamental class of S^j being the Thom-Pontrjagin class).

ii) an identification $(\hat{E},E) = (S^j \times D^{n-j}, S^j \times S^{n-j-1})$. The Poincaré decomposition of P then looks as follows :

$$\overline{P} = S^j \times D^{n-j} \cup P_+ , \qquad \partial P_+ = \partial P \amalg S^j \times S^{n-j-1},$$

and we suppose that $\alpha | \partial P = \text{id}$.

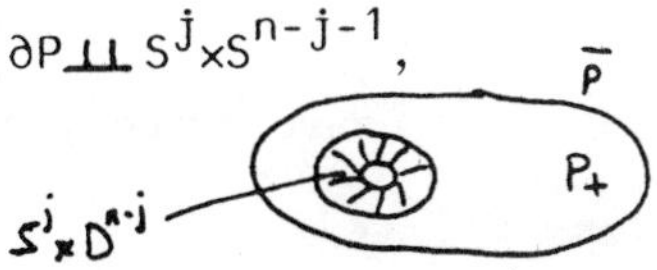

iii) a path $\gamma : [0,1] \longrightarrow P$ joining the base point of $S^j \times \{0\}$ to the base point of P such that the map $\alpha | S^j$ attached to the base point of P by $\alpha \cdot \gamma$ represents the class b in $\pi_j(P)$.

Form the PD-space R of formal dimension n+1 as follows :

$$R = P\times[0,1] \cup \mathcal{M}_\alpha \cup_{(S^j\times D^{n-j})} (D^{j+1}\times D^{n-j})$$

where $\mathcal{M}_\alpha$ is the mapping cylinder of $\alpha : \overline{P} \to P\times 1$. The map $f : P \longrightarrow X$ extends to $F : R \longrightarrow X$ by defining $F(u,t) = \overline{\beta}(u)$ for $(u,t) \in D^{j+1}\times D^{n-j}$.

Therefore R becomes a space over BG by defining $j^R = j^P \circ F$. Since α is a morphism of Poincaré spaces, the space $\mathcal{M}_\alpha$ is a Poincaré space (see the proof of (3.8)) with $(\mathcal{M}_\alpha)$ inducing $-(S^j\times D^{n-j})$ on $S^j\times D^{n-j}$. The class $(S^j\times D^{n-j})$ is the Thom-Pontrjagin class of $S^j\times D^{n-j}$, since it corresponds to the Thom-Pontrjagin class of S^j under the isomorphism μ of Section 3.C. Therefore, $(S^j\times D^{n-j})$ is induced from the Thom-Pontrjagin class $(D^{j+1}\times D^{n-j})$. Hence, by (3.4), there is a class (R) giving a Poincaré space structure on R, and (R,F) is a Poincaré cobordism between (P,f) and a pair (P',f').

The Poincaré space R (or the map F) is said to be obtained from P×[0,1] by attaching a **Poincaré handle** of index j+1 . The Poincaré space P' (or the map f') is said to be obtained from P by a **Poincaré surgery** of index j on the class b (or on the class $\overline{b}$).

(3.15) <u>**Proposition**</u> Let $\overline{b} \in \widetilde{K}_j(f)$ as above. Suppose that b is realizable by some PD-embedding (see (2.11)). If $j < n/2$, then a Poincaré surgery can be performed on $\overline{b}$.

The proof uses the following lemma :

(3.16) **Lemma** For $k \geq 3$, the homomorphism $\pi_r(BG_k) \longrightarrow \pi_r(BG)$ is bijective when $r < k$ and surjective when $r = k$.

Proof : This fact is classical for the map $BO_k \longrightarrow BO$ (it comes from the fibration $O_s \rightarrow O_{s+1} \rightarrow S^s$). We then use the fact that the following diagram :

$$\begin{array}{ccc} BO_k & \longrightarrow & BO \\ \downarrow & & \downarrow \\ BG_k & \longrightarrow & BG \end{array}$$

is (2k-2)-connected for $k \geq 3$ (see [Hr]). []

Proof of (3.15) : Let $(\nu,\overline{P},\alpha)$ be a PD-embedding of S^j into P (notations of (2.11)) realizing the class b (we also have a path γ joining the base point of $S^j \times 0$ to the base point of $\overline{P}$). We first prove that ν is trivial.

Since α is a homotopy equivalence, the pair $(j^{\overline{P}},(\overline{P}))$, where $j^{\overline{P}} = j^P \circ \alpha$ and $(\overline{P}) = \alpha_*^{-1}((P))$, constitutes a Spivak pair for $\overline{P}$. The map β permits us to produce a homotopy between $j^{\overline{P}}| S^j \times 0$ and a constant map. Since $j^{\overline{P}}| S^j \times 0$ classifies the join of ν with the Spivak bundle of S^j, we then have stable trivialization of ν. By Lemma (3.16), since $j < n-j$, this stable trivialization is induced by trivialization of ν. Chosing such a trivialization permits us to change $\overline{P}$ up to homotopy equivalence so that $\overline{P} = S^j \times D^{n-j} \cup P_+$. The same construction of the pair $(j^P,(P))$ as above makes P a Poincaré space and α a morphism of Poincaré spaces. However, this is not yet a Poincaré embedding of S^j into P, since the $\mathcal{H}$-fundamental class (P) might not induce the Thom-Pontrjagin class on S^j. Let $\eta \in \mathcal{H}_j(S^j)$ be the class such that $\mu^{-1}(\eta) \in \mathcal{H}_n(S^j \times D^{n-j})$ is the class induced by (P). The pair $(j^\nu + j^P | S^j, \eta)$ is a Spivak pair for S^j, as well as the standard pair $(*, (S^j))$. By [Br1, I.4.19], there is a fibre homotopy equivalence :

$$e : S^j \times S^N \longrightarrow S^j \times S^N$$

such that $T(e)_*((S^j)) = \eta$. The map e is characterized by an element of $\pi_j(BG)$. Since $j < n-j$, the homomorphism $\pi_j(BG_{n-j}) \to \pi_j(BG)$ is surjective. Therefore, the fibre homotopy equivalence e may be decomposed :

$$e : S^j \times S^{n-j-1} \times S^{N'} \xrightarrow{\bar{e} \times id} S^j \times S^{n-j-1} \times S^{N'}$$

Using the homotopy equivalence $\bar{e}$ (or rather its extension to the mapping cylinder of the projection $S^j \times S^{n-j-1} \to S^j$), one can change P by a homotopy equivalence so that the class induced by (P) on $S^j \times D^{n-j}$ is the Thom-Pontrjagin class. One has now a Poincaré embedding of S^j into P representing b, on which Poincaré surgery can be performed. [].

Proposition (3.15) together with the existence of smooth low-dimensional skeleta (Propositions (2.15), (2.16) and (2.17) permits us to perform low-dimensional Poincaré surgeries :

(3.17) **Proposition** Let X be a space over BG. Let $f : P \longrightarrow X$ be a map over BG, where P be Poincaré space of formal dimension n. Let $\bar{b} \in \widetilde{K}_j(P)$ with :

$$j = 0, \text{ if } n = 3, \quad \text{or}$$

$$j \leq 1, \text{ if } n = 4, \quad \text{or}$$

$$j \leq 2, \text{ if } n \geq 5.$$

Then, a Poincaré surgery can be performed on the class $\bar{b}$.

Proof : Let $\beta : S^j \longrightarrow P$ represent $\bar{b}$. By (2.15)-(2.17), the space P can be changed by a homotopy equivalence so that β is homotopic to a map into a codimension 0 submanifold of P. By general position in manifolds, β is homotopic to a manifold-embedding which, using its normal bundle, determines a PD-embedding of S^j into P. The result then follows from Proposition (3.15). []

We now turn our attention to Poincaré handle subtractions. Let (X,Y) be a pair over BG. Let P be Poincaré space of formal dimension n with a PD-decomposition of its boundary $\partial P = L_a \cup L_b$. Let $f : (P, L_b) \longrightarrow (X, Y)$. Let $\bar{b} \in \widetilde{K}_j(P, L_b)$ be represented by a commutative diagram

$$\begin{array}{ccc} (D^j, S^{j-1}) & \xrightarrow{\beta} & (P, L_b) \\ \cap & & \downarrow f \\ (D^j \times [0,1], S^{j-1} \times [0,1] \cup D^j \times \{1\}) & \xrightarrow{\overline{\beta}} & (X, Y) \end{array}$$

Let b be the class of β in $\pi_j(P,L_b)$. A **Poincaré handle** (of index j) realizing b consists of :

i) a Poincaré embedding $(\nu,\hat{E},\overline{P},\alpha,\text{etc})$ as described in (2.12) of (D^j,S^{j-1}) into (P,L_b), relative to L_a.

ii) an identification $(\hat{E},E) = (D^j \times D^{n-j}, D^j \times S^{n-j-1})$. The Poincaré decomposition $\overline{P} = \overline{P}_- \cup \overline{P}_+$, etc, then satisfies :

$$\overline{P}_- = D^j \times D^{n-j} \,,$$
$$\partial_- \overline{P} = S^{j-1} \times D^{n-j},$$

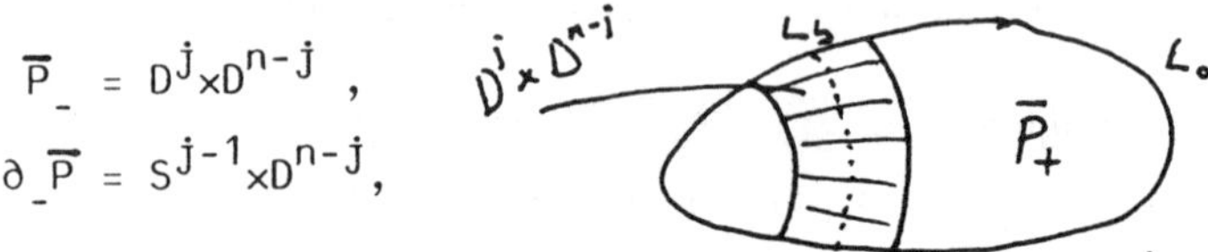

iii) a path $\gamma : [0,1] \longrightarrow P$ joining the base point of $D^j \times \{0\}$ to the base point of P such that the map $\alpha|(D^j,S^{j-1})$ attached to the base point of P by $\alpha \bullet \gamma$ represents b.

Let $P' = \overline{P}_+$. The boundary of P' is then Poincaré decomposed by $\partial P' = L_a \cup L'_b$, where L'_b is obtained from L_b by a surgery of index j-1 on $\beta | S^{j-1}$. The space P' is said to be obtained from P by a **Poincaré handle subtraction** (on the class b, relative to L_a). Observe that (P,L_b,L_a) and (P',L'_b,L_a) are Poincaré cobordant by the Poincaré cobordism $(W,V,L_a \times [0,1])$, where :

$W = \mathcal{M}_\alpha \quad , \quad V = \mathcal{M}_\alpha | L_b \cup P_-$

The map $\overline{\beta}$ permits us to extend $f : (P,L_b) \longrightarrow (X,Y)$ to $F : (W,V) \longrightarrow (X,Y)$, with $F(u,t) = f(u)$ for $u \in L_a$. The map $f' : F|(P',L_b')$ is said to be obtained from f by a **Poincaré handle subtraction** (on the class b, relative to L_a).

Poincaré handle subtraction are easier to deal with than surgeries. Indeed, a spherical bundle over D^j is always trivial, and there are no subtleties about Thom-Pontrjagin classes since there is a unique Poincaré space structure on the PD-space D^j (since $\mathcal{H}_*(D^j,\partial D^j) \equiv H_*(D^j,\partial D^j)$). Therefore, Lemma (3.14) permits us to prove the following

(3.18) **Proposition** Let (P,L_a,L_b,X,Y,f) be as above. Let $\overline{b} \in \widetilde{K}_j(P,L_b)$. Suppose that b can be realized by some PD-embedding relative to L_a. Then a Poincaré handle subtraction can be performed on $\overline{b}$, relative to L_a.

As in (3.17), the existence of smooth low-dimensional skeleta gives :

(3.19) **Proposition** Let (P,L_a,L_b,X,Y,f) be as above, with the formal dimension of P being $\geqslant 5$. Let $\overline{b} \in \widetilde{K}_j(P,L_b)$ with $j \leq 2$. Then a Poincaré handle subtraction can be performed on $\overline{b}$, relative to L_a

Finally, using (2.21), one obtains :

(3.20) **Proposition** Propositions (3.17) and (3.19) hold true in the category of finitely dominated PD-spaces.

4. THE POINCARE BORDISM GROUP AND THE EXACT SEQUENCE OF QUINN

A. PRESENTATION OF THE RESULTS

Let U be a CW-complex and V be a subcomplex of U. The classical theory of Thom and Atiyah asserts that the bordism group $\Omega_n(U,V)$ is isomorphic to the homology group $H_n(U,V;MO)$ (in the oriented case, ${}^{or}\Omega_n(U,V) = H_n(U,V;MSO)$). The bordism group ${}^{(or)}\Omega_n(U,V)$ is of course the group of bordism classes of maps $f : (M^n,\partial M) \longrightarrow (U,V)$, where M is a smooth compact (oriented) manifold.

<u>Question</u> : What happens when manifolds are replaced by PD-spaces ?

Let us define $\Omega_n^{(PD)}(U,V)$ to be the group of bordism classes of maps $f : (P,\partial P) \longrightarrow (U,V)$, where P is a PD-space of formal dimension n. Define ${}^{or}\Omega_n^{(PD)}(U,V)$ in the same way, requiring that each PD-space under consideration be oriented. Using the Spivak normal bundle it is possible to define, as in the Atiyah-Thom theory, a map from $\Omega_n^{(PD)}(U,V)$ into a homology group as follows : let $f : (P,\partial P) \longrightarrow (U,V)$ be a representative of an element of $\Omega_n^{(PD)}(U,V)$. The Spivak normal bundle theory (see [Br1, Section 1.4]) provides a spherical fibration $s_k^P : P \longrightarrow BG_k$, together with a "Thom-Pontrjagin class" $\alpha_k^P \in \pi_{n+k}(T(s_k^P),T(s_k^P|\partial P))$. The pair (s_k^P,α_k^P) is unique up to stable homotopy equivalence. With the notation of Section 3.A, there is the following diagram :

$$\begin{array}{ccc}
\hat{E}(s_k^P) & \xrightarrow{(f\bullet p)\times s_k^P} & U\times\hat{E}_k \\
\downarrow & & \downarrow \\
\hat{E}(s_k^P)/E(s_k^P)\cup\hat{E}(s_k^P|\partial P) & \longrightarrow & U\times\hat{E}_k \;/\; U\times E_k\cup V\times\hat{E}_k \\
\uparrow \alpha_k^P & & \downarrow \simeq \\
D^{n+k}/S^{n+k-1} & & (U/V)\wedge T(E_k)
\end{array}$$

($p : E(s_k^P) \longrightarrow U$ the projection of s_k^P)

This enables us to associate to (P,f) an element of $\pi_{n+k}((U/V)\wedge T(E_k))$. Passing to the limit over k gives a map $\Omega_n^{(PD)}(U,V)\longrightarrow H_n(U,V;MG)$ which is a homomorphism. In the oriented case, this homomorphism goes from ${}^{or}\Omega_n^{(PD)}(U,V)$ into $H_n(U,V;MSG)$.

The proof that the homomorphism $\Omega_n(U,V)\longrightarrow H_n(U,V;MO)$ is bijective relies on the transversality theory in manifolds. Such a transversality theory does not exist in Poincaré spaces and the map $\Omega_n^{(PD)}(U,V)\longrightarrow H_n(U,V;MG)$ is not an isomorphism. In [Qu], F. Quinn announced that this map fits in to a long exact sequence

$$\rightarrow L_n(\pi(U),\pi(V)) \longrightarrow {}^{or}\Omega_n^{(PD)}(U,V) \longrightarrow H_n(U,V;MSG) \longrightarrow$$

$$\longrightarrow L_{n-1}(\pi(U),\pi(V)) \longrightarrow \qquad (4.1)$$

involving the Wall surgery obstruction groups $L_n = L_n^h$. This sequence is essentially due to N. Levitt [Le3] when $X = pt$ and $Y = \emptyset$; the approach of L. Jones also gives such a sequence [Jo1], as well as the algebraic point of view of A. Ranicki [Ra2, p. 618]. F. Quinn did not publish any proof (for a critic and a partial realization of Quinn's program, see [Ht]). We give in this section a proof of Exact sequence (4.1), based on a new idea. See (4.5) for the precise statement. The corresponding statement for the non-orientable case, absent so far in the literature is also given in (4.6).

We shall consider a more general setting, matching with those of Section 3. Let (X,Y) be a pair over BG. The **Poincaré bordism group** $\Omega_n^P(X,Y)$ is defined as the group of bordism classes of pairs (P,f), where P is a Poincaré space and $f : (P,\partial P) \longrightarrow (X,Y)$ is a map over BG. The group law is the disjoint union. We denote by $\Omega_n^P(X)$ the group $\Omega_n^P(X,\emptyset)$. There is a long exact sequence :

$$(4.2) \rightarrow \Omega_n^P(Y) \longrightarrow \Omega_n^P(X) \longrightarrow \Omega_n^P(X,Y) \longrightarrow \Omega_{n-1}^P(Y) \longrightarrow \Omega_{n-1}^P(X) \rightarrow$$

As for the obvious homomorphism $\Omega_n^P(X,Y) \longrightarrow \Omega_n^Q(X,Y)$, it fits in to the long exact sequence :

$$(4.3)\quad \to \Omega^P_n(X,Y) \longrightarrow \Omega^Q_n(X,Y) \longrightarrow \Omega^{QP}_n(X,Y) \longrightarrow$$
$$\longrightarrow \Omega^P_{n-1}(X,Y) \longrightarrow \Omega^Q_{n-1}(X,Y) \to$$

in which the relative bordism groups $\Omega^{QP}_n(X,Y)$ are defined as follows : the elements of $\Omega^{QP}_n(X,Y)$ are bordism classes of pairs (Q,f), where :

1) Q is a Q-space of formal dimension n with a Q-decomposition $\partial Q = Q_1 \cup P$,
2) $f : (Q,Q_1) \longrightarrow (X,Y)$ is a map over BG,
3) the Q-space P is a Poincaré space.

Exact sequences like (4.2) exist also for the groups $\Omega^Q_n(X,Y)$ and for the groups $\Omega^{QP}_n(X,Y)$, with the obvious commutative braid diagrams also involving (4.3).

Let X be a space over BG and denote by $\pi(X)$ its fundamental groupoid [Sp, Section 1.7]. Recall that a map $w^{BG} : BG \longrightarrow \mathbf{R}P^\infty$ inducing an isomorphism on the fundamental groups was chosen in Section 3.A, so X becomes a space over $\mathbf{R}P^\infty$ by $w^X = w^{BG} \circ j^X$. This provides a homomorphism of groupoids :

$$w^X : \pi(X) \longrightarrow \{\pm 1\}$$

Therefore, if (X,Y) is a pair over BG, we may consider the Wall group $L_n(\pi(X),\pi(Y);w^X)$ (in this section, $L_n = L^h_n$) defined as in [Wa2]. (Strictly speaking, Wall defines these groups when X and Y have finitely many connected components; we then take $L_n(\pi(X),\pi(Y);w^X)$ to be the inductive limit of

$L_n(\pi(X_i),\pi(Y_i);w^X i)$, where (X_i,Y_i) run over all the subpairs of (X,Y) with X_i and Y_i having finitely many connected components.) If X is connected, one may replace $\pi(X)$ with the fundamental group $\pi_1(X)$ and w^X is considered as a homomorphism from $\pi_1(X) \longrightarrow \{\pm 1\}$.

We are now in position to state the main result of this section :

(4.4) **Theorem** Let (X,Y) be a pair over BG. For all n, there is a natural homomorphism

$$\sigma_n \ : \ \Omega^{QP}_{n+1}(X,Y) \longrightarrow L_n(\pi(X),\pi(Y);w^X)$$

$$(\sigma_n \ : \ \Omega^{QP}_{n+1}(X) \longrightarrow L_n(\pi_1(X);w^X)$$

if X is connected and $Y = \emptyset$)

satisfying the following properties :

a) The following diagram commutes :

$$\begin{array}{ccccccc}
\longrightarrow & \Omega^{QP}_{n+1}(X,Y) & \longrightarrow & \Omega^{QP}_{n}(Y) & \longrightarrow \\
 & \downarrow \sigma_n & & \downarrow \sigma_{n-1} & \\
\longrightarrow & L_n(\pi(X),\pi(Y);w^X) & \longrightarrow & L_{n-1}(\pi(Y);w^Y) & \longrightarrow
\end{array}$$

$$\begin{array}{ccccccc}
\longrightarrow & \Omega^{QP}_{n}(X) & \longrightarrow & \Omega^{QP}_{n}(X,Y) & \longrightarrow \\
 & \downarrow \sigma_{n-1} & & \downarrow \sigma_{n-1} & \\
\longrightarrow & L_{n-1}(\pi(X);w^X) & \longrightarrow & L_{n-1}(\pi(X),\pi(Y);w^X) & \longrightarrow
\end{array}$$

b) If $Y = \emptyset$ or $n \geqslant 4$, then σ_n is injective.

c) The image of σ_n depends only on $(\pi(X),\pi(Y);w^X)$.

d) σ_n is surjective for $n \geqslant 5$ in general and for $n \geqslant 4$ if $Y = \emptyset$.

e) If X is connected and $Y = \emptyset$, then $\mathrm{Im}_2 = L_2(0) = \mathbb{Z}/2\mathbb{Z}$.

f) $\Omega_n^{QP}(X,Y) = 0$ if $n \leqslant 1$.

As a corollary, we obtain the exact sequence mentioned above :

(4.5) **Corollary** Let U be a connected CW-complex and V a subcomplex of U. Then, there is an exact sequence :

$$\rightarrow {}^{or}\Omega_n^{(PD)}(U,V) \longrightarrow H_n(U,V;MSG) \longrightarrow L_{n-1}(\pi_1(U),\pi(V)) \longrightarrow$$

$$\longrightarrow {}^{or}\Omega_{n-1}^{(PD)}(U,V) \rightarrow \cdots \longrightarrow H_5(U,V;MSG) \longrightarrow L_4(\pi_1(U),\pi(V)).$$

In addition, when $V = \emptyset$, this sequence may be prolonged to

$$\longrightarrow H_5(U;MSG) \longrightarrow L_4(\pi_1(U)) \longrightarrow {}^{or}\Omega_4^{(PD)}(U) \longrightarrow$$

$$\longrightarrow H_4(U;MSG) \longrightarrow L_3(\pi_1(U)).$$

Proof : We apply Theorem (4.4) to the case

$$X = U\times BSG, \quad Y = V\times BSG, \quad j^X : U\times BSG \xrightarrow{\text{proj}} BSG \longrightarrow BG.$$

Use (3.10) to obtain the isomorphism $\Omega_n^Q(U\times BSG, V\times BSG) \cong H_n(U,V;MSG)$. As for the isomorphism $\mu : \Omega_n^P(U\times BSG,V\times BSG) \xrightarrow{\cong} {}^{or}\Omega_n^{(PD)}(U,V)$, it is given by $\mu(P,f,(P)) \longmapsto (P,\bar{f})$, where $\bar{f}$ is the composition $P \xrightarrow{f} U\times BSG \xrightarrow{\text{proj}} U$. The surjectivity of μ holds since $(P,\bar{g}) = \mu(P,(\bar{g},j^P),(P))$, where $(j^P,(P))$ is a

Spivak pair for P. The uniqueness of Spivak pairs [Br1, Theorem I.4.19] guarantees the injectivity of μ. []

Replacing BSG by BG gives the non-orientable case :

(4.6) **Corollary** Let U and V as in (4.5), with U and V connected. There is an exact sequence :

$$\to \Omega_n^{(PD)}(U,V) \longrightarrow H_n(U,V;MG) \longrightarrow$$
$$\longrightarrow L_{n-1}(\pi_1(U)\times\{\pm 1\},\pi_1(V)\times\{\pm 1\};w) \longrightarrow \Omega_{n-1}^{(PD)}(U,V)\to$$
$$\ldots \longrightarrow H_5(U,V;MG) \longrightarrow L_4(\pi_1(U)\times\{\pm 1\},\pi_1(V)\times\{\pm 1\};w),$$

where $w : \pi_1(U)\times\{\pm 1\} \longrightarrow \{\pm 1\}$ is the natural projection. If $V = \emptyset$, the exact sequence may be prolonged :

$$\to H_5(U;MG) \longrightarrow L_4(\pi_1(U)\times\{\pm 1\};w) \longrightarrow \Omega_4^{(PD)}(U) \longrightarrow$$
$$\longrightarrow H_4(U;MG) \longrightarrow L_3(\pi_1(U)\times\{\pm 1\};w).$$

The proof of Theorem (4.4) will follow from Propositions (4.7) and (4.8) below:

(4.7) **Proposition** Let (X,Y), $\tilde{j}^X : X \longrightarrow BO$ be a pair over BO. For all n, there is a homomorphism

$$\sigma_n(\tilde{j}^X) \; : \; \Omega_{n+1}^{QP}(X,Y) \longrightarrow L_n(\pi(X),\pi(Y);w^X)$$
$$(\sigma_n(j^X) \; : \; \Omega_{n+1}^{QP}(X) \longrightarrow L_n(\pi_1(X);w^X)$$
if X is connected and $Y = \emptyset$)

which is natural (in the category over BO) and which satisfies Properties a) to f) of Theorem (4.4).

(4.8) **Proposition** Let (X,Y) be a pair over BG. If (X,Y) is 2-connected, then $\Omega_n^{QP}(X,Y) = 0$ for all n.

Assuming Proposition (4.7) and (4.8) (which will be proved in Section 4.B and 4.C), the proof of Theorem (4.4) proceeds as follows : let (X,Y) be a pair over BG. Replace the inclusion $BO \to BG$ by a Serre fibration i. The pull-back diagram

$$\begin{array}{ccc} (\bar{X},\bar{Y}) & \xrightarrow{\bar{j}^{X}} & BO \\ \downarrow & & \downarrow i \\ (X,Y) & \xrightarrow{j^{X}} & BG \end{array}$$

gives rise to the commutative diagram

$$\begin{array}{ccccccccc} \Omega_n^{QP}(\bar{Y}) & \longrightarrow & \Omega_n^{QP}(\bar{X}) & \longrightarrow & \Omega_n^{QP}(\bar{X},\bar{Y}) & \longrightarrow & \Omega_{n-1}^{QP}(\bar{Y}) & \longrightarrow & \Omega_{n-1}^{QP}(\bar{X}) \\ \downarrow \beta_Y & & \downarrow \beta_X & & \downarrow \beta_{XY} & & \downarrow \beta_Y & & \downarrow \beta_X \\ \Omega_n^{QP}(Y) & \longrightarrow & \Omega_n^{QP}(X) & \longrightarrow & \Omega_n^{QP}(X,Y) & \longrightarrow & \Omega_{n-1}^{QP}(Y) & \longrightarrow & \Omega_{n-1}^{QP}(X) \end{array}$$

As the map $BO \to BG$ is 2-connected, so are the pairs $(X,\bar{X})$ and $(Y,\bar{Y})$. By Proposition (4.8) the homomorphisms β_X and β_Y are isomorphisms, and so is β_{XY} by the Five Lemma. Using Proposition (4.7), one is able to define the homomorphism σ_n as the composed homomorphism

$$\begin{array}{ccccc} \Omega_{n+1}^{QP}(X,Y) & \xrightarrow{\beta_{XY}^{-1}} & \Omega_{n+1}^{QP}(\bar{X},\bar{Y}) & \xrightarrow{\sigma_n(\bar{j}^{X})} & L_n(\pi(\bar{X}),\pi(\bar{Y});w^{\bar{X}}) \\ & & & & \downarrow \cong \\ & & & & L_n(\pi(X),\pi(Y);w^{X}) \end{array}$$

This homomorphism σ_n has all the properties listed in Theorem (4.4) since, according to Proposition (4.7) this is the case for $\sigma_n(\tilde{j}^X)$. This ends the proof of Theorem (4.4), provided we establish Propositions (4.7) and (4.8).

The definition of σ_n has the following consequence :

(4.9) **Proposition** Let (X,Y) be a pair over BG and let $\tilde{j}^X : X \to BO$ be a lifting of j^X. Then, the homomorphisms $\sigma_n(\tilde{j}^X)$ and σ_n of Proposition (4.7) and Theorem (4.4) are equal. In particular, $\sigma_n(\tilde{j}^X)$ depends only on j^X.

Proof : The lifting $\tilde{j}^X$ gives rise to a section $\gamma : (X,Y) \longrightarrow (\bar{X},\bar{Y})$ such that $\tilde{j}^{\bar{X}} \circ \gamma = \tilde{j}^X$. Therefore, one has the following commutative diagram :

$$\begin{array}{ccc}
\Omega^{QP}_{n+1}(X,Y) & \xrightarrow{\sigma_n(\tilde{j}^X)} & L_n(\pi(X),\pi(Y);w^X) \\
\downarrow \gamma_* & & \downarrow L\gamma \\
\Omega^{QP}_{n+1}(\bar{X},\bar{Y}) & \xrightarrow{\sigma_n(\tilde{j}^{\bar{X}})} & L_n(\pi(\bar{X}),\pi(\bar{Y});w^{\bar{X}})
\end{array}$$

As γ is a section, one has $\beta_{XY} \circ \gamma_* = \mathrm{id}$. As seen above, β_{XY} is an isomorphism, so is γ_*. $L\gamma$ is also an isomorphism, being a right inverse of the isomorphism $L_n(\pi(\bar{X}),\pi(\bar{Y});w^{\bar{X}}) \xrightarrow{\cong} L_n(\pi(X),\pi(Y);w^X)$. Therefore, $\sigma_n(\tilde{j}^X) = \sigma_n$. []

B. PROOF OF PROPOSITION (4.7)

Let (X,Y) be a pair over BG and $\tilde{j}^X : X \to BO$ be a lifting of j^X. These data are fixed throughout the proof and the symbol $\tilde{j}^X$ will be most of the time omited in the notation. The homomorphism $\sigma_n = \sigma_n(\tilde{j}^X)$ will be a composed homomorphism

$$\Omega^{QP}_{n+1}(X,Y) \longrightarrow \Omega^P_n(X,Y) \xrightarrow{\bar{\sigma}_n} L_n(\pi(X),\pi(Y);w^X) .$$

To define $\bar{\sigma}_n$ $(= \bar{\sigma}_n(\tilde{j}^X))$, let (P,f) represent an element of $\Omega^P_n(X,Y)$ and let ξ be the stable vector bundle over P with characteristic map $\tilde{j}^X \circ f : P \to BO$. The pair $(\xi,(P))$ determines a surgery problem

$$\begin{array}{ccc} \nu_M & \longrightarrow & \xi \\ \downarrow & & \downarrow \\ (M,\partial M) & \xrightarrow{g} & (P,\partial P) \end{array}$$

with associated surgery obstruction $\sigma(g) \in L_n(\pi(P),\pi(\partial P);w^P)$. We define $\bar{\sigma}_n(P,f)$ as $f_*(\sigma(g))$. For this definition to make sense, we have to check that $f_*(\sigma(g))$ does not depend on the choice of (P,f). We give here the details for the case $Y = \emptyset$ and leave the general case to the reader (it is the same principle but requires more notation).

Let (P',f') be another representative of the class of (P,f) in $\Omega^P_n(X)$. There exists then a Poincaré cobordism (W,P,P') and a map $F : W \to X$ extending f and f'. There is also a compatibility condition on the classes(P) and (P') which implies that

the corresponding surgery problems

extends to a surgery problem

$$\begin{array}{ccc} \nu_S & \longrightarrow & \xi_W \\ \downarrow & & \downarrow \\ S & \xrightarrow{G} & W \end{array}$$

The surgery obstruction $\sigma(G)$ lies in the Wall group $L_{n+1}(\pi(W),\pi(P \cup P');w^W)$. Let us consider the following diagram :

$$\begin{array}{ccc} L_{n+1}(\pi(W),\pi(P \cup P');w^W) & & \\ \downarrow \delta & & \\ L_n(\pi(P \cup P');w^{(P \cup P')}) & \xleftarrow[\cong]{\beta} & L_n(\pi(P);w^P) \oplus L_n(\pi(P');w^{P'}) \\ \downarrow & & \\ L_n(\pi(W);w^W) & & \end{array}$$

where the vertical column is exact. Then, $\delta(\sigma(G)) = \beta(\sigma(g),-\sigma(g'))$. This implies that $\sigma(g)$ and $\sigma(g')$ have same image in $L_n(\pi(W);w^W)$, and then $f_*(\sigma(g)) = f'_*(\sigma(g'))$. Observe that this argument only involves the definition of the surgery obstruction and is then valid for all n.

As said before, a similar argument shows that $\bar{\sigma}_n : \Omega_n^P(X,Y) \longrightarrow L_n(\pi(X),\pi(Y);w^X)$ is well defined. The homomorphism $\bar{\sigma}_n$ defined this way satisfies Property a) of Theorem (4.4) (using the last sentence of Theorem (3.2) of [Wa2]). It is then enough to prove Parts b)-e) of (4.7) when $Y = \emptyset$; the remaining statements are consequences of the five Lemma applied to the diagram of Part a). (Property f) will be shown at the end of this section). As one has the following commutative diagram :

$$\begin{array}{ccccc} \Omega^{QP}_{n+1}(X) & \longrightarrow & \Omega^{P}_{n}(X) & \longrightarrow & L_n(\pi(X);w^X) \\ \uparrow & & \uparrow & & \uparrow \\ \bigoplus_i \Omega^{QP}_{n+1}(X_i) & \longrightarrow & \bigoplus_i \Omega^{P}_{n}(X_i) & \longrightarrow & \bigoplus_i L_n(\pi(X_i);w^{X}i) \end{array}$$

(X_i = connected component of X)

it is sufficient to establish Parts b)-e) of (4.7) when X is connected. The relevant Wall group is then $L_n(\pi_1(X);w^X)$. In (3.11), we have proved that the lifting $\tilde{j}^X$ provides an isomorphism $\Omega^M_n(X) \xrightarrow{\simeq} \Omega^Q_n(X)$. This isomorphism obviously factors through $\Omega^P_n(X)$. This splits Exact Sequence (4.3) and gives decompositions

$$\Omega^P_n(X) \quad = \quad \Omega^{QP}_{n+1}(X) \oplus \Omega^Q_n(X) \quad = \quad \Omega^{QP}_{n+1}(X) \oplus \Omega^M_n(X) \qquad (4.10)$$

(such a splitting occurs only when j^X admits a lifting to BO). This implies that the injectivity of σ_n is equivalent to the condition $\ker\bar{\sigma}_n \subset \Omega^M_n(X)$. On the other hand, the above factorization allows us to consider $\Omega^{QP}_{n+1}(X)$ as a quotient $\Omega^{QP}_{n+1}(X) \cong \Omega^P_n(X)/\Omega^M_n(X)$ and σ_n is the homomorphism induced on this quotient by $\bar{\sigma}_n$. Therefore, the surjectivity of σ_n and $\bar{\sigma}_n$ are equivalent.

<u>Injectivity of $\overline{\sigma}_n$ for $n \geq 3$</u> : Let (P,f) represent an element of $\Omega_n^P(X)$ whose image under $\overline{\sigma}_n$ is equal to zero. Let

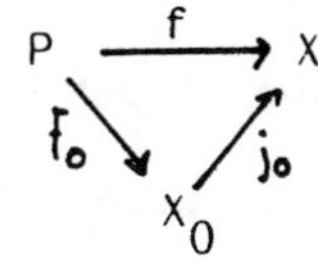

be a surgery problem associated as above to $(\tilde{j}^X \circ f,(P))$, with surgery obstruction $\sigma(g) \in L_n(\pi_1(P);w^P)$. By hypothesis, one has $f_*(\sigma(g)) = 0$. Let X_0 be a connected space with finitely presented fundamental group such that there is a factorization

$$\begin{array}{ccc} P & \xrightarrow{f} & X \\ & {}_{f_0}\searrow \quad \nearrow_{j_0} & \\ & X_0 & \end{array}$$

with $(f_0)_*(\sigma(g)) = 0$ in $L_n(\pi_1(X_0);w^{X_0})$. If $n \geq 4$, one may perform low dimensional surgeries on P and M (using (3.17)) in order to transform (g,f,f_0) into $(g',f'.f_0')$

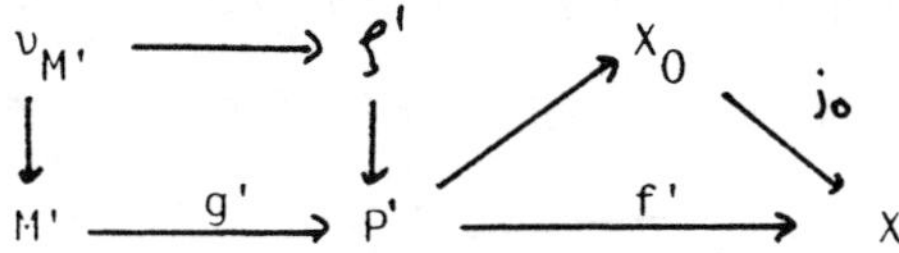

with $(f_0)_*(\sigma(g')) = 0$ and $\pi_1(P') = \pi_1(X_0)$. Then $\sigma(g') = 0$. Suppose first that $n \geq 5$. Then there is a cobordism $(W,M'M")$ and an extension of g' to a normal map $G : W \longrightarrow P'$ such that $g" = G|M" : M" \longrightarrow P'$ is a homotopy equivalence. The space $R = W \cup \mathcal{M}_{g"}$ is a Poincaré cobordism from (P',f') to

$(M'', f' \circ g'')$. The later represents an element of $\Omega_n^M(X)$; this proves that $\ker \bar{\sigma}_n \subset \Omega_n^M(X)$ which, by what was said above, proves the injectivity of σ_n for $n \geq 5$.

When $n = 4$, the same argument applies after replacing P' and M' by their connected sum with a suitable number of copies of $S^2 \times S^2$'s [CS2, Theoerem 2.1]. This can be done without changing the class of (P,F) in $\Omega_4^P(X)$.

If $n = 3$ in the above argument, we can only achieve $\pi_1(P') \to \pi_1(X_0)$ surjective. Odd dimensional surgery [Wa2, Chapter 6] can be used to construct W and g'' but g'' will only be a a $\mathbb{Z}\,\pi_1(X_0)$-homology equivalence. The space R is, a priori, only a $\mathbb{Z}$-$\pi_1(X_0)$-Poincaré space. But, by low-dimensional surgery, $f' \circ G : W \longrightarrow X_0$ is cobordant, rel ∂W, to $F : W \longrightarrow X_0$ so that $\pi_1 F$ is an isomorphism. On the other hand, g'' being a degree one map between PD spaces, $\pi_1 g''$ is onto. Let $R = W \cup \mathcal{M}_{g''}$. One has the following commutative diagram :

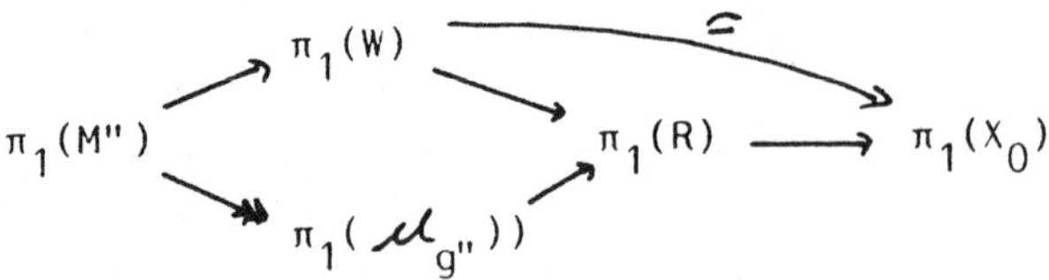

By Van-Kampen's theorem, one deduces that $\pi_1(W) \longrightarrow \pi_1(R)$ is onto and then $\pi_1(R) \cong \pi_1(X_0)$. Therefore, R is a Poincaré cobordism between M' and P', which proves, as in the case $n \geq 5$, the injectivity of σ_3.

Injectivity of σ_2 and computation of its image : Let (P,f) represent an element of $\Omega_2^P(X)$. Then, P has the homotopy type of a closed manifold by [EM]. Therefore, the normal bordism classes of P (in the sense of [Wa2, Chapter 10]) are in natural bijection with $[P,G/O] = \pi_2(G/O) \xrightarrow[\cong]{a} L_2(0) = Z/2Z$, where a is the Arf invariant. This proves that $\bar{\sigma}_2$ (and therefore σ_2) is injective with image $L_2(0) \subset L_2(\pi_1(X);w^X)$.

Surjectivity of σ_n for $n \geq 4$: Let $u \in L_n(\pi_1(X);w^X)$. There exists a homomorphism $r : T \longrightarrow \pi_1(X)$, where T is finitely presented and such that $u = r_*(u_1)$, for some $u_1 \in L_n(T;w^X \circ r)$. Let K^2 be a finite 2-dimensional complex with $\pi_1(K) = T$. Choose a map $\beta : K \longrightarrow X$ with $\pi_1\beta = r$.

Let ν be the stable vector bundle over K with characteristic map $\tilde{j}^X \circ \beta : K \longrightarrow BO$ and denote by $(-\nu)$ the stable bundle over K such that $\nu \oplus (-\nu)$ is trivial. We first construct a compact manifold V^n which collapses onto K and which has ν as stable normal bundle (or equivalently $(-\nu)$ as stable tangent bundle). (For $n \geq 5$, V will be the stable thickening of K with stable tangent bundle $(-\nu)$ [Wa1, Proposition 5.1].) One starts with the n-disk $B = B^n$. Attach a set of 1-handles to ∂B in bijection with the 1-cells of K and then a set of 2-handles in bijection with the 2-cells of K so that V has the simple homotopy type of K. The 1-handle representing a generator z of $\pi_1(K)$ is attached in an orientable way if $\pi_1(j^X \circ \beta)(z) = 1$ and in a non-orientable way otherwise. Each 2-handle is attached in order to get $(-\nu)$ as stable tangent bundle for V. This is possible since $\pi_1(SO_{n-2}) \to \pi_1(SO)$ is onto if $n \geq 4$.

Suppose first that $n \geq 6$. Let us consider the normal map

$$\begin{array}{ccc} \nu_{\hat{V}} & \longrightarrow & \nu_{\bar{V}} \\ \downarrow & & \downarrow \\ \hat{V} & \longrightarrow & \bar{V} \end{array}$$

where $\bar{V} = V \cup (\partial V \times [0,1])$ (V union a collar), $\hat{V} = V \cup W$, $g|V$ = id and

$$g|W : (W,\partial V,\partial V) \longrightarrow (\partial V\times[0,1],\partial V\times\{0\},\partial V\times\{1\})$$

is the $([n/2]-1)$-connected normal map with surgery obstruction u_1 constructed in [Wa2, Theorem 5.8 and 6.5]. (We consider u_1 $L_n(\pi_1(\partial V);w^X \bullet r)$ using the isomorphism $\pi_1(\partial V) = \pi_1(V) = \pi_1(X)$.) Thus $\sigma(g) = u_1$ and $g|\partial\hat{V} : \partial\hat{V} \to \partial\bar{V}$ is a homotopy equivalence.

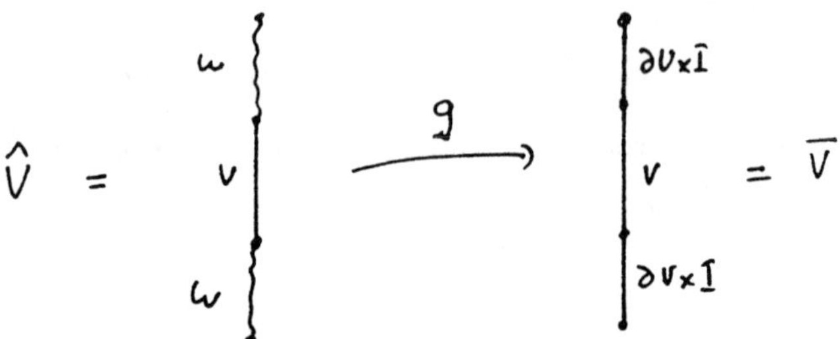

Form the Poincaré space $P^n = \hat{V} \cup_{g|\partial V} \bar{V}$. There is a map

$F : P \longrightarrow X$ such that $F|\overline{V} = f$ and $F|\hat{V} = f \circ g$. Let $\hat{V}_1$ and $\hat{V}_2$ denote two copies of $\hat{V}$ and $\hat{V}_1 \cup \hat{V}_2$ be the manifold obtained by glunig $\hat{V}_1$ to $\hat{V}_2$ by $\mathrm{id} : \partial\hat{V}_1 \rightarrow \partial\hat{V}_2$. There is a normal map

$$\begin{array}{ccc} \nu(\hat{V}_1 \cup \hat{V}_2) & \xrightarrow{b_G} & \zeta \\ \downarrow & & \downarrow \\ \hat{V}_1 \cup \hat{V}_2 & \xrightarrow{G} & P \end{array}$$

where ζ is the stable vector bundle whith charactic map $\hat{j}^X \circ F$. The map G is defined as $G|\hat{V}_1 = \mathrm{id} : \hat{V}_1 \longrightarrow \overline{V}$ and $G|\hat{V}_2 = g : \hat{V}_2 \longrightarrow \overline{V}$. The existence of such a normal map proves that $j^X \circ F$ classifies the Spivak bundle of P and that (P,f) represents an element of $\Omega^P_n(X)$, where $(P) = Tb_G((\hat{V}_1 \cup \hat{V}_2))$. By construction, one has

$$\overline{\sigma}_n(P,f) = F_*(\sigma(G)) = f_*(\sigma(g)) = f_*(u_1) = u.$$

When $n = 5$, the same argument applies as well. Indeed, the normal map $g|W : W \longrightarrow \partial V \times [0,1]$ with surgery obstruction u_1 does exist provided one adds to V, along ∂V, a large enough collection of trivial 2-handles. This uses the "special case of Theorem 3.1" of [CS2, p. 502].

Another modification of the above argument gives the surjectivity of σ_4. Let T, u_1 and V^4 defined as above. The only difference is that the homomorphism $\pi_1(\partial V) \longrightarrow \pi_1(V) = T$ is now only surjective. Let $\Gamma_4 = \Gamma_4(Z\pi_1(\partial V) \longrightarrow ZT)$ be the Cappell-Shaneson surgery obstruction group [CS1], and let u_2 be an element of Γ_4 going to u_1 under the surjection $\Gamma_4 \twoheadrightarrow$

$L_4(T;w^X \circ r)$. One can construct the normal map $g : \hat{V} \longrightarrow \bar{V}$ as above, with $g|W$ being the normal map with Cappell-Shaneson surgery obstruction u_2, obtained as in [CS1, Theorem 1.8]. Here, the map $f|\partial\hat{V} : \partial\hat{V} \longrightarrow \partial\bar{V}$ is only a $\mathbb{Z}T$-homology equivalence and therefore, the space P is, a priori, just a $\mathbb{Z}T$-Poincaré space. However, let us determine $\pi_1(P)$ by the Van-Kampen theorem. We are dealing with the following push-out diagram :

$$\begin{array}{ccc} \pi_1(\partial\hat{V}) & \xrightarrow{i} & \pi_1(\hat{V}) \\ \downarrow & & \downarrow \\ \pi_1(\bar{V}) & \xrightarrow{j} & \pi_1(P) \end{array}$$

The homomorphism i is surjective since, by construction of V and W, the manifold V has only 0, 1 and 2 handles. Therefore, the homomorphism j is also surjective. The homomorphism $\pi_1 g$: $\pi_1(\hat{V}) \longrightarrow \pi_1(\bar{V})$ produces homomorphism $e : \pi_1(P) \longrightarrow \pi_1(\bar{V})$ which is a section of j. As $\pi_1 g$ is onto (g is a map of degree one between manifolds), the map e is onto, which implies that j is an isomorphism. Thus $\pi_1(P) \cong \pi_1(V) \cong T$ and P is a Poincaré space. As above, one constructs $(P,f) \in \Omega_4^P(X)$ such that $\sigma_4(P,f) = u$.

<u>Proof that</u> $\mathrm{Im}\,\sigma_3$ <u>depends only on</u> $(\pi_1(X);w^X)$: Let us consider the diagram

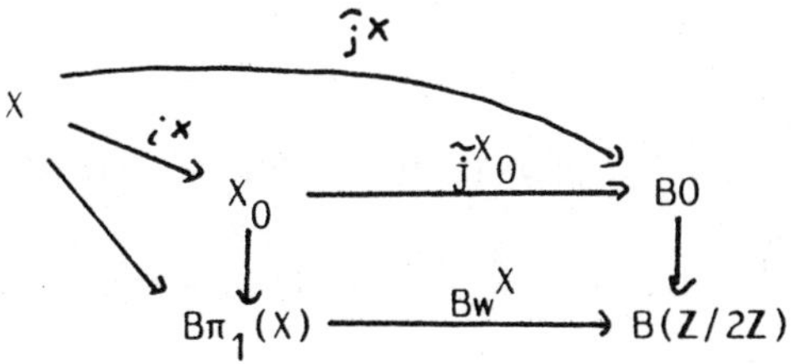

where the square diagram is pull-back. we shall prove that the homomorphism $\Omega_4^{QP}(X) \longrightarrow \Omega_4^{QP}(X_0)$ is surjective. Therefore, using the isomorphism $L_3(\pi_1(X);w^X) \cong L_3(\pi_1(X_0);w^{X_0})$ (since $\pi_1(X) \cong \pi_1(X_0)$), one obtains a natural identification of $\mathrm{Im}\sigma_3^{(X)}$ with $\mathrm{Im}\sigma_3^{(X_0)}$. This will prove our asssertion, since the couple (X_0, j^{X_0}) only depends on the couple $(\pi_1(X), w^X)$.

Using the splitting

$$\Omega_3^P(X) \cong \Omega_4^{QP}(X) \oplus \Omega_3^M(X)$$

(and the same for X_0), it is enough to prove that the composed homomorphism

$$\Omega_3^P(X) \longrightarrow \Omega_3^P(X_0) \longrightarrow \Omega_3^P(X_0)/\Omega_3^M(X_0)$$

is surjective. Let (P,F) represent an element of $\Omega_3^P(X_0)$, and let $g : M \longrightarrow P$ be a normal map of degree one in the normal bordism class determined by $(j^{X_0},(P))$. By (2.17) and (3.3), there is a Poincaré decomposition of P :

$$P = P_1 \cup D^3 , \quad P_1 \cap D^3 = \partial P_1 = \partial D^3 = S^2$$

Putting g transverse regular to the center of D^3 gives a manifold decomposition

$$M = M_1 \cup (\amalg D^3) , \; M_1 \cap (\amalg D^3) = \partial M_1 = \partial(\amalg D^3)$$

respected by g and so that $g|\amalg D^3 = \amalg$ diffeomorphism. If D_-, $D_+ \subset \amalg D^3$ are so that $g|D_\pm$ is of degree ± 1, then $D_- \cup D_+$ may be removed from $g^{-1}(D^3)$ by the following procedure : join D_- to D_+ by an arc γ so that $g \circ \gamma$ is null-homotopic (this is possible since $\pi_1 g : \pi_1(M) \longrightarrow \pi_1(P)$ is onto, since g is of degree one). Let D be the connected sum of D_- with D_+, using a tube along γ. As $g \circ \gamma$ is null-homotopic, g can be changed by a homotopy so that, in $g^{-1}(D^3)$, $D_- \cup D_+$ is replaced by D. But $g|D$ is of degree zero. A second homotopy will then remove D from $g^{-1}(D^3)$. As g is of degree one, this procedure will eventually end up with a manifold decomposition $M = M_1 \cup D^3$, respected by g and $g|D^3$ is a diffeomorphism. We claim that there is a lifting $\tilde{f}_1$:

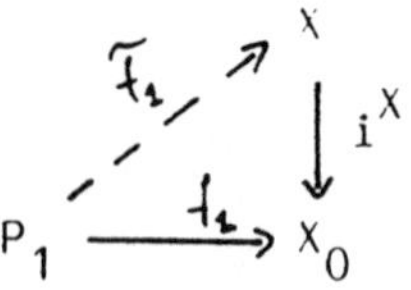

of $f_1 = f|P_1$. Indeed, the first and only one obstruction to the existence of $\tilde{f}_1$ is an element $t \in H^2(P_1;\pi_1(F))$, where F is the theoretical fiber of i^X. If the homomorphism $\pi_2 i^X$ is onto, then $\pi_1(F) = 0$. If not, as $\pi_2(X_0) = \pi_2(BO) = Z_2$, the map i^X admits a lifting $\hat{i}^X : X \longrightarrow \hat{X}_0$, where $\hat{X}_0$ is the fiber of $w_2 : \hat{X}_0 \longrightarrow K(Z_2,2)$ (w_2 is the second Stiefel-Whitney class of the bundle with characteristic map $j^X 0$). Then $t = f^*(w_2) = 0$, since the second Stiefel-Whitney class of the Spivak bundle of a Poincaré space of formal dimension 3 vanishes (a classical consequence of the Wu's formula).

Form the PD-space $R = P_1 \cup M_1$, using the gluing map $g|\partial M_1$, and make R a Poincaré space by (3.4). Define $h : R \longrightarrow X$ by $h|P_1 = f_1$ and $h|M_1 = f_1 \circ g$. The pair (R,h) represents an element of $\Omega_3^P(X)$ whose image in $\Omega_3^P(X_0)$ is equal to $(P,f) - (M, f \circ g)$. The surjectivity of the composed homomorphism

$$\Omega_3^P(X) \longrightarrow \Omega_3^P(X_0) \longrightarrow \Omega_3^P(X_0)/\Omega_3^M(X_0)$$

is then established.

Proof of Property f) : As j^X admits a lifting $\tilde{j}^X$ through BO, the homomorphism $\Omega_n^M(X,Y;\tilde{j}^X) \longrightarrow \Omega_n^Q(X,Y)$ is an isomorphism by Proposition (3.11). Using the factorization

$$\begin{array}{ccc} \Omega_n^M(X,Y;\tilde{j}^X) & \xrightarrow{\ \simeq\ } & \Omega_n^Q(X,Y) \\ & \searrow \quad \nearrow & \\ & \Omega_n^P(X,Y) & \end{array}$$

it is sufficient to prove that $\Omega_n^P(X,Y) \longrightarrow \Omega_n^Q(X,Y)$ is injective for $n \leq 1$. But this follows from the easy fact that $\Omega_n^M(X,Y;\tilde{j}^X) \longrightarrow \Omega_n^P(X,Y)$ is surjective for $n \leq 1$.

C. PROOF OF PROPOSITION (4.8)

Throughout this section, (X,Y) will denote a 2-connected pair over BG.

a) PLAN OF THE PROOF

Let us consider a sequences of spaces over BG as follows :

$$\begin{array}{c} B_0 \subset B_1 \subset B_2 \subset \cdots \\ \searrow \quad \searrow \downarrow \quad \cdots \\ BG \end{array}$$

with the following properties :

1) B_0 is the space BO

2) $B_{i+1} = B_i \cup$ (one cell of dimension $j(i)$), with $j(i+1) \geqslant j(i) \geqslant 3$.

3) The map $\lim B_i \longrightarrow BG$ is a homotopy equivalence.

Such a sequence exists, the condition $j(i) \geqslant 3$ can be fulfilled since the map $BG \rightarrow BO$ is 2-connected.

We shall prove the following assertions on the homomorphism $t_n : \Omega_n^P(X,Y) \longrightarrow \Omega_n^Q(X,Y)$ (as said at the begining, (X,Y) is a 2-connected pair over BG) :

Assertion $A^e(i,n)$: If j^X admits a lifting $\tilde{j}^X : X \rightarrow B_i$, then the homomorphism t_n is surjective.

Assertion $\mathbf{A}^m(i,n)$: If j^X admits a lifting $\tilde{j}^X : X \to B_i$, then the homomorphism t_n is injective.

Assertion $\mathbf{A}(i,n)$: If j^X admits a lifting $\tilde{j}^X : X \to B_i$, then the homomorphism t_n is an isomorphism.

It is convenient to recapitulate these assertions as follows : we consider on the set **N×N** the reverse lexicographic order, namely

$$(i,n) \prec (i',n') \text{ if either } n < n' \text{ or } (n = n' \text{ and } i < i').$$

Assertion $\mathbf{RA}(i,n)$: If $(j.k) \preceq (i,n+1)$, and if j^X admits a lifting $\tilde{j}^X : X \to B_j$ then $\Omega_k^{QP}(X,Y) = 0$.

Assertion $\mathbf{RA}(i,n)$ is then equivalent to $\mathbf{A}(j,k)$ for $(j,k) \preceq (i,n)$ and $\mathbf{A}^e(j,k)$ for $(j,k) \preceq (i,n+1)$.

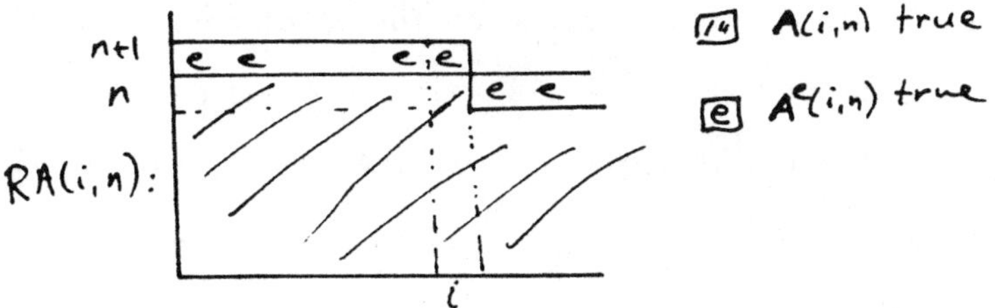

Assertion $\mathbf{RA}(i,n)$ for all i and n implies Proposition (4.8) in the following way :

<u>Lemma</u> The following statements are equivalent

a) the statement of proposition (4.8)

b) For any integer m, there is $n \geq m$ such that $\mathbf{RA}(0,n)$ is true.

Proof : Only b) implies a) requires a proof. Condition b) clearly implies that **RA**(i,n) is true for all (i,n). Consider the map $B_i \to BG$ as a Serre fibration and form the pull-bak diagram

$$\begin{array}{ccc} (X_i,Y_i) & \longrightarrow & B_i \\ \downarrow & & \downarrow \\ (X,Y) & \longrightarrow & BG \end{array}$$

The pairs (X,X_i) and (Y,Y_i) have the same connectivity as the pair (BG,B_i). This connectivity tends to infinity when $i \to \infty$. Let $n \in \mathbf{N}$. As Ω^Q_* is a homology theory over BG, there exists an integer i such that the homomorphism $\Omega^Q_k(X_i,Y_i) \longrightarrow \Omega^Q_k(X,Y)$ is an isomorphism for $k \leq n$. If i is large enough, the same will be true for the homomorphism $\Omega^P_k(X_i,Y_i) \longrightarrow \Omega^P_k(X,Y)$. Indeed, if P is a Poincaré space of formal dimension n, the pair $(P,\partial P)$ is of cohomological dimension n and one can use obstruction theory. Therefore $\Omega^{QP}_n(X_i,Y_i) = \Omega^{QP}_n(X,Y)$. If i is large enough, the pair (X_i,Y_i) is again 2-connected and, by Assertion **RA**(i,n), one has $\Omega^{QP}_n(X_i,Y_i) = 0$. Therefore $\Omega^{QP}_n(X,Y) = 0$. []

Assertion **RA**(0,n) will be proven by induction on n. The plan of the proof is to use special argument in low dimensions and a general induction argument which works in dimension ≥ 9. More precisely :

1) using Proposition (4.7) one establishes **A**(0,n) for all n. Therefore, to prove that **RA**(0,n) implies **RA**(0,n+1), it is enough to prove that **RA**(i,n) implies **RA**(i+1,n).

2) general properties of Ω_*^{QP} prove **RA**(0,3).

3) we prove that **RA**(i,n) implies **RA**(i+1,n) for $n \leq 5$. The proof uses the fact that the elements of $\Omega_n^P(X,Y)$, for $n \leq 5$, have manifold representatives.

4) to extend the argument of 3) in higher dimensions, we introduce an additional induction assertion, (Assertion π(i,n), which is a special case of a Poincaré (π-π)-theorem. One has

- **RA**(i,n) impies π(i,n)
- **RA**(i-1,n) and π(i,n) imply **RA**(i,n).

Using that a Poincaré space of formal dimension 6, 7 or 8 with $(P,\partial P)$ 2-conneted has a Poincaré decomposition into two manifolds, one proves π(i,n) for all i when n = 6,7,8. This establishes **RA**(0,9).

5) we are finally able to prove that **RA**(i-1,n) implies **RA**(i,n) for $n \geq 9$. This is done by introducing a last induction hypothesis, **S**(i,n), which is essentially the surgery abilities requested to establish π(i,n). Assertion **S**(i,n) is proved along with **RA**(i,n) (they are actually put together in a new assertion **RAS**(i,n)). Therefore, **RA**(0,n)

is true for arbitrary large n's and, by the previous lemma, Proposition (4.8) is proved.

b) PROOF OF **A**(0,n) AND **RA**(0,3)

(4.10) **Lemma** Let (X,Y) be a k-connected pair over BG. Then $\Omega^{Q}_{n}(X,Y) = \Omega^{QP}_{n}(X,Y) = 0$ for $n \leq k$ and $\Omega^{P}_{n}(X,Y) = 0$ for $n \leq k-1$.

Proof : The equality $\Omega^{Q}_{n}(X,Y) = 0$ for $n \leq k$ holds since Ω^{Q}_{*} is a homology theory over BG (use Spectral sequence (3.13)). The equality $\Omega^{P}_{n}(X,Y) = 0$ for $n \leq k-1$ comes from the fact that if P is a Poincaré space of formal dimension n, then the pair $(P,\partial P)$ is of cohomology dimension n. Therefore, by obstruction theory, any map $f : (P,\partial P) \longrightarrow (X,Y)$ is, if $n < k$, homotopic relative to ∂P, to a map from P into Y. The equality $\Omega^{QP}_{n}(X,Y) = 0$ for $n \leq k$ then comes from Exact Sequence (4.3).

(4.11) **Lemma** A(0,n) is true for all n.

Proof : As $B_0 = BO$, Proposition (4.7) provides us a commutative diagram :

$$\begin{array}{ccc} \Omega^{QP}_{n+1}(Y) & \longrightarrow & \Omega^{QP}_{n+1}(X) \\ \downarrow \sigma_n(\tilde{j}^Y) & & \downarrow \sigma_n(\tilde{j}^X) \\ L_n(\pi_1(Y);w^Y) & \longrightarrow & L_n(\pi_1(X);w^X) \end{array}$$

By Properties b) and c) of (4.4), the homomorphisms $\sigma_n(\tilde{j}^Y)$ and $\sigma_n(\tilde{j}^X)$ are injective and have same image. Therefore $\Omega^{QP}_{n+1}(Y) \cong \Omega^{QP}_{n+1}(X)$ for all n. []

(4.11.b) **Corollary** To prove that **RA**(0,n) implies **RA**(0,n+1), it is enough to prove that **RA**(i,n) implies **RA**(i+1,n). []

A direct consequence of (4.10) and (4.11) is

(4.11.c) **Corollary** **RA**(0,3) is true. []

c) PROOF OF RA(i-1,n) implies RA(i,n) FOR $n \leq 5$

(4.12) **Lemma** For all (i,n), **A**(i-1,n-1) and $\mathbf{A}^e(i-1,n)$ imply $\mathbf{A}^e(i,n)$.

Proof : Chose a lifting $\tilde{j}^X : X \to B_i$ of j^X. One has $B_i = B_{i-1} \cup e^j$, where e^j is a cell of dimension $j = j(i-1) \geq 3$. Write $e^j = e^{j-1} \times [-1,1]$. Replace $\tilde{j}^X : X \to B_i$ by a Serre fibration and set

$$X_- = (\tilde{j}^X)^{-1}(B_{i-1} \cup (e^{j-1} \times [-1,0]))$$

$$X_+ = (\tilde{j}^X)^{-1}(B_{i-1} \cup (e^{j-1} \times [0,1]))$$

$$X_- \cap X_+ = (j^X)^{-1}(B_{i-1} \cup (e^{j-1} \times \{0\}))$$

Replacing the inclusion $Y \subset X$ by a Serre fibration, the decomposition $X = X_- \cup X_-$ gives a corresponding decomposition $Y = Y_- \cup Y_+$. Observe that the usual inclusions between the spaces $Y_- \cap Y_+$, $X_- \cap X_+$, Y_-, Y_+, X_-, X_+ are 2-connected and that $j^X | X_-$ and $j^X | X_+$ admit liftings through B_{i-1}.

Let (Q,f) represent an element of $\Omega_n^Q(X,Y)$. Use the classical trick to replace f by a Serre fibration (still called f) and set $Q_\pm = f^{-1}(X_\pm)$, $\partial_\pm Q = f^{-1}(Y_\pm)$. One obtains a decomposition

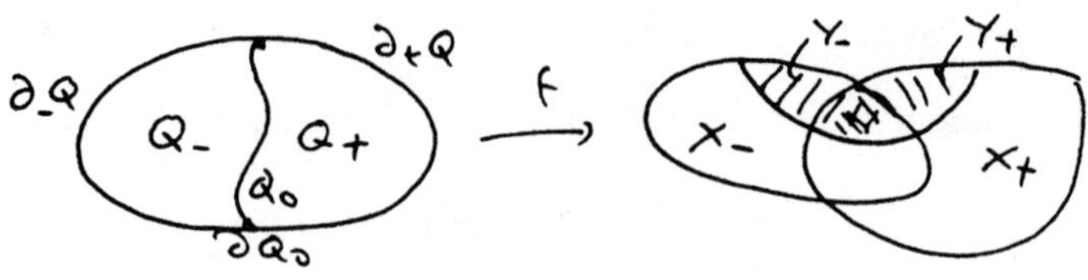

$$Q = Q_- \cup Q_+ \ , \quad Q_- \cap Q_+ = Q_0$$

$$\partial Q = \partial_- Q \cup \partial_+ Q \ , \quad \partial_- Q \cap \partial_+ Q = \partial Q_0$$

This decomposition can be made a Q-decomposition by (3.4). We consider of course that $\partial Q_{\pm} = \partial_{\pm} Q \cup Q_0$ and $\partial(\partial_{\pm} Q) = \partial Q_0$. The map $f|\partial_- Q : (\partial_- Q, \partial Q_0) \longrightarrow (Y_-, Y_- \cap Y_+)$ represents an element of $\Omega^Q_{n-1}(Y_-, Y_- \cap Y_+)$. As $j^X|Y_-$ factors through B_{i-1}, the homomorphism $\Omega^P_{n-1}(Y_-, Y_- \cap Y_+) \longrightarrow \Omega^Q_{n-1}(Y_-, Y_- \cap Y_+)$ is surjective by **A**(i-1,n-1). Therefore, there is a Q-cobordism $W(\partial_- Q)$ from $\partial_- Q$ to $\overline{\partial_- Q}$ (restricting to $W(\partial Q_0)$, from ∂Q_0 to $\overline{\partial Q_0}$) so that $\overline{\partial_- Q}$ is a Poincaré space with $\partial(\overline{\partial_- Q}) = \overline{\partial Q_0}$, and f extends to $F : (W(\partial_- Q), W(\partial Q_0)) \longrightarrow (Y_-, Y_- \cap Y_+)$. The map $F : (Q_- \cup W(\partial_- Q), Q_0 \cup W(\partial Q_0)) \longrightarrow (X_-, Y_-)$ represents an element of $\Omega^{QP}_n(X_-, Y_-)$. The latter vanishes by **Assertions** A(i-1,n-1) and $\mathbf{A}^e$(i-1,n). Hence there exists a Q-cobordism $W(Q_- \cup W(\partial_- Q))$, relative to $\overline{\partial_- Q}$, from $Q_- \cup W(\partial_- Q)$ to a Poincaré space $\overline{Q}_-$ and $W(Q_- \cup W(\partial_- Q))$ contains in its boundary a Q-cobordism $W(Q_0 \cup W(\partial Q_0))$ from $Q_0 \cup W(\partial Q_0)$ to a Poincaré space $\overline{Q}_0$. Applying the same procedure again to

$$F : (Q_+ \cup W(Q_0 \cup W(\partial Q_0)), \partial_+ Q \cup W(\partial Q_0)) \longrightarrow (X_+, Y_+)$$

(one has $\Omega_n^{QP}(X_+,Y_+) = 0$) finishes the construction of a Q-cobordim between (Q,f) and an element of $\Omega_n^P(X,Y)$. []

(4.12.b) **Corollary** Suppose that **RA**$(i-1,n)$ is true. Then **RA**(i,n) is true if and only if $\mathbf{A}^m(i,n)$ is true.

Proof : If **RA**$(i-1,n)$ is true, **RA**(i,n) will be true if $\mathbf{A}^m(i,n)$ and $\mathbf{A}^e(i,n+1)$ are true. **RA**$(i-1,n)$ implies $\mathbf{A}^e(i,n)$ by Lemma (4.12). []

(4.13) **Lemma** **RA**$(i-1,n)$ implies **RA**(i,n) if $n \leqslant 5$.

Proof : By (4.12.b), it is enough to establish $\mathbf{A}^m(i,n)$. By Corollary (4.11.c), one may assume that $n \geqslant 2$.

Suppose that j^X admits a lifting $\tilde{j}^X : X \longrightarrow B_i$. Consider the pull-back diagram

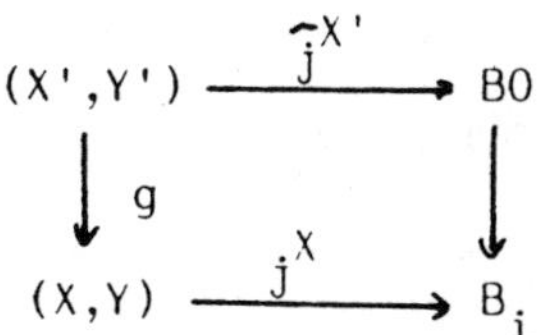

One has a commutative diagram

$$\begin{array}{ccc} \Omega_n^P(X',Y') & \xrightarrow{\ t'_n\ } & \Omega_n^Q(X',Y') \\ \downarrow r_n^P & & \downarrow r_n^Q \\ \Omega_n^P(X,Y) & \xrightarrow{\ t_n\ } & \Omega_n^Q(X,Y) \end{array}$$

Let $q \in \ker t_n \cap \operatorname{Im} r_n^P$. Write $q = r_n^P(P,f)$, where (P,f) represents an element of $\Omega_n^P(X',Y')$. The pair (X',Y') is 2-connected and $\tilde{j}^{X'}$ admits the lifting $\bar{j}^{X'}$ through BO. Therefore, Proposition (3.11) and Assertion $\mathbf{A}(0,n)$ imply that the following homomorphisms

$$\begin{array}{ccc} \Omega_n^M(X',Y';\bar{j}^{X'}) & \longrightarrow & \Omega_n^Q(X',Y') \\ & \searrow \quad \Omega_n^P(X',Y') \quad \nearrow t_n' & \end{array}$$

are all isomorphisms. Hence, we may assume that P is a manifold and (P) its Thom-Pontrjagin class.

As $t_n(q) = 0$, there exists a Q-space Q of formal dimension n+1, with a Q-decomposition $\partial Q = P \cup N$ ($\partial P = \partial N$), and an extension $\bar{f} : (Q,N) \longrightarrow (X,Y)$ of f. As P is a manifold, the decompositioins $X = X_- \cup X_+$, $Y = Y_- \cup Y_+$ appearing in the proof of (4.12) induce, by transversality, corresponding decompositions $P = P_- \cup P_+$, $P_- \cap P_+ = P_0$, $\partial P = \partial_- P \cup \partial_+ P$, $\partial_- P \cap \partial_+ P = \partial P_0$, so that all the spaces under consideration are manifolds. This decomposition extends to a Q-decomposition of (Q,N). The same argument as for the proof of Lemma (4.12) (using $\mathbf{A}(i-1,n-1)$ and $\mathbf{A}^e(i-1,n+1)$), shows that (Q,N) is Q-cobordant (relative to P) to Poincaré spaces. Therefore $q = 0$ and we have shown that $\ker t_n \cap \operatorname{Im} r_n^P = \{0\}$ (for all n).

To finish up the proof of (4.13), it is enough to prove that

r^P_n is surjective when $n \leq 5$. Let (P,f) represents an element of $\Omega^P_n(X,Y)$. By low dimensional Poincaré surgeries, using (3.17), the pair $(P,\partial P)$ can be made $(n-3)$-connected (since $n \leq 5$, $n-3 \leq 2$). The obstructions to obtaining a lifting :

$$\begin{array}{ccc} & & X' \\ & \tilde{f} \nearrow & \downarrow \\ P & \xrightarrow{f} & X \end{array}$$

lie in $H^i(P,\pi_{i-1}(F)) = H_{n-i}(P,\partial P;\pi_{i-1}(F))$, where F is the theoretical fibre of $X' \longrightarrow X$. As $\pi_{i-1}(F) = 0$ for $i \leq 3$ and $(P,\partial P)$ is $(n-3)$-connected, all the bstruction groups are zero. Then, $\tilde{f} : (P,\partial P) \rightarrow (X',Y')$ does exist and one has $r^P_n(P,\tilde{f}) = (P,f)$. []

Using (4.11.b), (4.11.c) and (4.13), one deduces

(4.14) **Corollary** **RA**(0,6) is true.

d) ASSERTION $\Pi(i,n)$ AND PROOF OF **RA**(0,9)

We here introduce a new assertion.

Assertion $\Pi(i,n)$: Let $g : (R,\partial R) \longrightarrow (P,\partial P)$ be a morphism of Poincaré spaces (R and P of formal dimension n). Suppose that the pair $(P,\partial P)$ is 2-connected and that j^P admits a lifting through B_i. Then g is Poincaré bordant to id_P.

Assertion $\Pi(i,n)$ is a provisional statement of the $(\pi$-$\pi)$-theorem (for the definitive statement, see (5.2)).

Observe that the map g of Assertion $\Pi(i,n)$ is Q-bordant to id_P. Therefore, $\Pi(i,n)$ is implied by $\mathbf{A}^m(i,n)$. On the other hand, one has

(4.15) **Lemma** $\mathbf{RA}(i-1,n)$ and $\Pi(i,n)$ imply $\mathbf{RA}(i,n)$.

Proof : By Corollary (4.12.b), it sufices to prove $\mathbf{A}^m(i,n)$. Let (P,f) represent an element $q \in \Omega^P_n(X,Y)$. By low dimensional Poincaré surgeries, using (3.17), the pair $(P,\partial P)$ can be made 2-connected (we may assume $n \geqslant 6$, in view of (4.14)).

We first proceed as in the proof of Lemma (4.12). If j^X admits a lifting through B_i, we split f over $B_{i-1} \cup e^j \times\{0\}$, obtaining a Q-decomposition of P

$$P = P_- \cup P_+ \ , \quad P_- \cap P_+ = P_0$$

$$\partial P = \partial_- P \cup \partial_+ P \ , \quad \partial_- P \cap \partial_+ P = \partial P_0$$

and $j^P{\pm}$ admits a lifting through B_{i-1}. Use $\mathbf{A}(i-1,n-1)$ and $\mathbf{A}^e(i-1,n)$ to show that id_P is Q-cobordant to $g : (R,\partial R) \to (P,\partial P)$, where R is a Poincaré space which is Poincaré decomposed :

$$R = R_- \cup R_+ \ , \quad R_- \cap R_+ = R_0$$

$$\partial R = \partial_- R \cup \partial_+ R \ , \quad \partial_- R \cap \partial_+ R = \partial R_0$$

By Assertion $\Pi(i,n)$, the map g is Poincaré cobordant to id_P and then q has a representative with underlying map $h = f \circ g : (R,\partial R) \to (X,Y)$.

If $t_n(q) = 0$, there exists a Q-space Q with a Q-decomposition $\partial Q = R \cup N$ and an extension $H : (Q,N) \longrightarrow (X,Y)$ of h. By surgery on Q-spaces (see (3.8)), one can assume that the pair (Q,N) is 2-connected. Use the classical trick to replace H by a Serre fibration

$$\begin{array}{ccc} Q & \xrightarrow{H} & X \\ \simeq \cap & \nearrow & \\ \hat{Q} & \hat{H} & \text{Serre fibration} \end{array}$$

Set $Q_{\pm} = H^{-1}(X_{\pm})$. Replace the inclusion $N \subset Q \subset \hat{Q}$ by a Serre fibration $\hat{N} \to \hat{Q}$ and consider it an an inclusion (using, as usual the mapping cylinder). We get this way a corresponding decomposition $\hat{N} = \hat{N}_- \cup \hat{N}_+$ and the usual inclusions between the spaces $\hat{N}_- \cap \hat{N}_+$, $\hat{Q}_- \cap \hat{Q}_+$, $\hat{N}_{\pm}$, $\hat{Q}_{\pm}$, $\hat{N}$ and $\hat{Q}$ are 2-connected. The inclusion $R \subset Q \subset \hat{Q}$ decomposes into inclusions $R_{\pm} \subset \hat{Q}_{\pm}$, etc. In other words, we have extended the Poincaré decomposition of R into a Q-decomposition of $\hat{Q}$.

Now, the same argument as in the proof of Lemma (4.12), using $\mathbf{A}(i-1,n)$ and $\mathbf{A}^e(i-1,n+1)$, permits us to replace $\hat{Q}$, up to Q-cobordism and relative to R, by a Poincaré space. Thus q = 0 and this establishes $\mathbf{A}^m(i,n)$. []

To prove Assertion $\pi(i,n)$, we shall often deal with the following situation : let $f : (P,\partial P) \longrightarrow (Q,\partial Q)$ be a morphism of Q-spaces, where P is a Poincaré space of formal dimension n. Suppose that one has Q-decompositions of Q and P :

$P = P_- \cup P_+ \ , \quad P_- \cap P_+ = P_0$

$\partial P = \partial_- P \cup \partial_+ P \ , \quad \partial_- P \cap \partial_+ P = \partial P_0$

$Q = Q_- \cup Q_+ \ , \quad Q_- \cap Q_+ = Q_0$

$\partial Q = \partial_- Q \cup \partial_+ Q \ , \quad \partial_- Q \cap \partial_+ Q = \partial Q_0$

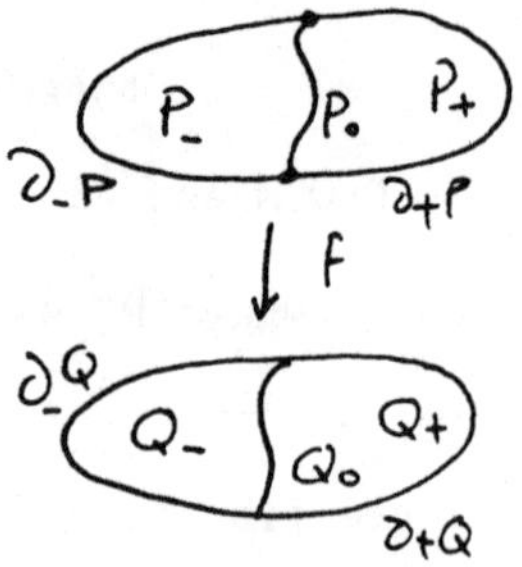

which are respected by f. As in the surgery setting, we introduce the (homology) **kernels** K_j of f which are defined as follows :

$$K_j(P_0) = H_{j+1}(f : P_0 \longrightarrow Q_0; Z\pi)$$

$$K_j(P_0, \partial P_0) = H_{j+1}(f : (P_0, \partial P_0) \longrightarrow (Q_0, \partial Q_0); Z\pi)$$

$$K_j(P_\pm) = H_{j+1}(f : P_\pm \longrightarrow Q_\pm; Z\pi) \qquad (\pi = \pi_1(Q))$$

$$K_j(P_\pm, \partial_\pm P) = H_{j+1}(f : (P_\pm, \partial_\pm P) \longrightarrow (Q_\pm, \partial_\pm Q); Z\pi)$$

$$K_j(P) = H_{j+1}(f : P \longrightarrow Q; Z\pi)$$

$$K_j(P, \partial P) = H_{j+1}(f : (P, \partial P) \longrightarrow (Q, \partial Q); Z\pi)$$

Using the usual Mayer-Vietoris argument in homology, one easily obtains the following lemma :

(4.16) <u>**Lemma**</u> One has the following (Mayer-Vietoris) exact sequence :

$$\longrightarrow K_j(P_-) \oplus K_j(P_+) \longrightarrow K_j(P) \longrightarrow K_{j-1}(P_0) \longrightarrow K_{j-1}(P_-) \oplus K_{j-1}(P_+) \longrightarrow$$

and

$$\longrightarrow K_j(P_-, \partial_- P) \oplus K_j(P_+, \partial_+ P) \longrightarrow K_j(P, \partial P) \longrightarrow K_{j-1}(P_0, \partial P_0) \longrightarrow$$
$$\longrightarrow K_{j-1}(P_-, \partial_- P) \oplus K_{j-1}(P_+, \partial_+ P) \longrightarrow$$

To establish Assertion $\pi(i,n)$, a special argument is first required for $6 \leq n \leq 8$.

(4.17) **Lemma** $\pi(i,n)$ is true for all i if n = 6, 7 or 8.

Proof : The proof is independant of i. Let P be a Poincaré space of formal dimension n such that the pair $(P,\partial P)$ is 2-connected. Compose the map $j^P : P \to BG$ with the map $BG \to B(G/O)$ [MM]. Let us consider the map $\Sigma(G/O) \to B(G/O)$ which is adjoint to $id_{B(G/O)}$ (using that $\Omega B(G/O) \cong G/O$). The fiber of this map is the joint $G/O*G/O$ (see [BM]). As the first non-zero homotopy group of G/O is $\pi_2(G/O) = Z_2$, one has $\pi_j(G/O*G/O) = 0$ for $j \leq 4$ and $\pi_5(G/O*G/O) = Z_2$. The first obstruction to find a lifting

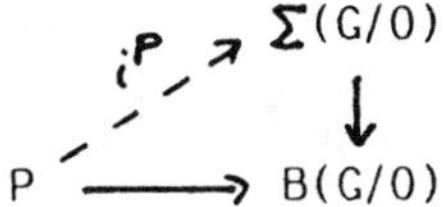

belongs to $H^6(P;Z_2) = H_{n-6}(P,\partial P;Z_2)$. If $n \leq 8$ and $(P,\partial P)$ is 2-connected, this obstruction group is zero as well as the higher ones. Therefore the lifting i^P exists. Replace i^P by a Serre fibration and set $P_0 = (i^P)^{-1}(G/O)$, where G/O lies in $\Sigma(G/O)$ in the obvious way. We obtain a decomposition $P = P_- \cup P_+$, etc, of P, where $P_\pm$ are the preimages of the two cones of $\Sigma(G/O)$. This decomposition can be made a Q-decomposition by (3.3). The maps $j^P\pm$ admit liftings $\bar{j}^P\pm : P_\pm \to BO$. Observe that the spaces G/O, $\Sigma(G/O)$, ∂P and P have

The homotopy type of complexes with finite skeleta. By Proposition (A.2) of the appendix at the end of this book, one deduces that the spaces $P_{\pm}$, P_0, ∂P_0, and $\partial_{\pm}P$ are all homotopy equivalent to complexes with finite skeleta.

Let $g : (R,\partial R) \longrightarrow (P,\partial P)$ be a map as in the statement of $\Pi(i,n)$, with $n \geqslant 6$. Make the pair $(R,\partial R)$ 2-connected by low dimensional Poincaré surgeries, using (3.17). As above, replace the map g by a Serre fibration and set $R_{\pm} = g^{-1}(P_{\pm})$, etc. One thus gets a Q-decomposition analoguous to those of P, which is respected by g. The lifting $\tilde{j}^{R}_{-}|R_0 : R_0 \to BO$ corresponds to a normal map of degree one $h_0 : (V_0^-,\partial V_0^-) \longrightarrow (R_0,\partial R_0)$, where $(V_0^-,\partial V_0^-)$ is a manifold pair.

Case $n = 6$: Perform surgeries in dimension 0, 1 and 2 on ∂V_0^- and V_0^- so that $\pi_1(\partial V_0^-) \xrightarrow{\cong} \pi_1(\partial R_0)$ and $\pi_1(V_0^-) \xrightarrow{\cong} \pi_1(R_0)$. By construction, the pair $(R_0,\partial R_0)$ is 2-connected and then $\pi_1(\partial V_0^-) \cong \pi_1(V_0^-)$. Digging out 2-handles on $(V_0^-,\partial V_0^-)$ (as in [Wa2, p. 40]), one can obtain that $K_j(V_0^-,\partial V_0^-) = 0$ for $j \leq 2$. The map $\tilde{j}^{R}_{+} \circ h_0 : V_0^- \to BO$ produces a normal map of degree one $r : V_0^+ \to V_0^-$. The classical proof of the $(\pi$-$\pi)$-theorem in the odd dimensional case [Wa2, Chapter 4], which is valid for $n \geqslant 7$, may be worked out in dimension 5 as well, using the stable 4-dimensional surgery theory [CS2]. This proves that r is normally cobordant to a homotopy equivalence after adding on V_0^- and V_0^+ a convenient number of trivial 2-handles (on their boundaries). Thus, we can assume that r is a homotoopy equivalence.

Make r transverse regular to some point $x_- \in \partial V_0^-$ and chose a point x_+ in the preimage of x_- so that r restricted to some neighbourhood D_+ of x_+ in V_0^+ is an orientation preserving diffeomorphism from D_+ onto a neighbourhood D_- of x_- into V_0^-. One may suppose that $D_\pm$ is the image of an embedding of $(D^4\times[0,1],D^4\times\{0\}) \hookrightarrow (\partial V_0^\pm,V_0^\pm)$. We call $D_\pm$ the **based disk** of $V_0^\pm$.

The lifting $\tilde{j}^R-$ permits us to extend $h_0 : V_0^- \longrightarrow R_0$ to a normal map of degree one $h_- : V_- \longrightarrow R_-$. With the lifting $\tilde{j}^R+$, one can extend $h_0 \circ r : V_0^+ \longrightarrow R_0$ to a a normal map of degree one $h_+ : V_+ \longrightarrow R_+$. This gives us a morphism of Q-spaces $h : (V,\partial V) \longrightarrow (R,\partial R)$, where $V = V_- \cup_r V_+$. The space V is a Poincaré space since r is a homotopy equivalence. Set $\partial V_\pm = \partial_\pm V \cup V_0^\pm$.

As $V_\pm$ are manifolds, one can perform surgeries (away of V_0) so that one obtains a morphism of Q-spaces $\bar{h} : \bar{V} = \bar{V}_- \cup_r \bar{V}_+ \longrightarrow R$ such that

a) $\bar{h}$ induces an isomorphism on the fundamental group of all the spaces under considedration ($\bar{V}_\pm$, $\partial_\pm\bar{V}$, V_0, and ∂V_0)

b) $K_j(\bar{V}_\pm) = K_j(\bar{V}_\pm,\partial_\pm\bar{V}) = 0$ for $j \leq 2$.

The Mayer-Vietoris sequences of (4.16) show that $K_j(\bar{V}) = K_j(\bar{V},\partial\bar{V}) = 0$ for $j \leq 2$. As both $\bar{V}$ and R are Poincaré spaces and $\bar{h}$ is of degree one, one has, according to [Wa2, Chapter

2], the following isomorphisms

$$K_j(\bar{V}) = K^{6-j}(\bar{V},\partial\bar{V}) \quad \text{and} \quad K_j(\bar{V},\partial\bar{V}) = K^{6-j}(\bar{V})$$

where $K^j(\bar{V}) = H^{j+1}(\bar{h} : \bar{V} \longrightarrow R; Z\pi)$, etc ($\pi = \pi_1(R)$). As $\pi_1\bar{h}$ is an isomorphism, Condition b) above implies that $K^j(\bar{V}) = K^j(\bar{V},\partial\bar{V}) = 0$ for $j \leq 2$. Therefore, for $j \geq 4$, one has $K_j(\bar{V}_\pm) = = 0 = K_j(\bar{V}_\pm,\partial_\pm\bar{V})$ and the only kernels which might not vanish are those of the following exact sequence :

$$0 \to K_3(\partial\bar{V}) \to K_3(\bar{V}) \to K_3(\bar{V},\partial\bar{V}) \to K_2(\partial\bar{V}) \to 0$$

Moreover, $K_3(\bar{V})$ and $K_3(\bar{V},\partial\bar{V})$ are stably $Z\pi$-free [Wa2, Lemma 2.3]. By replacing V by its boundary connected sum with copies of $S^3 \times D^3$ (connected sum made along, say, ∂_+V), one can assume that $K_3(\bar{V},\partial\bar{V})$ is $Z\pi$-free with basis $e_1,\ldots,e_k$.

As $K_2(V_0^-,\partial V_0^-) = 0 = K_2(V_0^+,\partial V_0^+)$ by construction, the Mayer-Vietoris sequences (4.16) imply that the homomorphism

$$K_3(\bar{V}_-,\partial_-\bar{V}) \oplus K_3(\bar{V}_+,\partial_+\bar{V}) \longrightarrow K_3(\bar{V},\partial\bar{V})$$

is surjective. Select elements

$$(e_i^-,e_i^+) \in K_3(\bar{V}_-,\partial_-\bar{V}) \oplus K_3(\bar{V}_+,\partial_+\bar{V})$$

having image e_i is $K_3(\bar{V},\partial\bar{V})$. The spaces $\bar{V}_+$ and $\bar{V}_-$ are manifolds and the pairs $(\bar{V}_-,\partial_-\bar{V})$ and $(\bar{V}_+,\partial_+\bar{V})$ are 2-connected. Therefore, the elements $e_i^\pm$ may be represented by disjoint 3-handles $\mathcal{E}_i^\pm : (D^3 \times D^3, S^2 \times D^3) \hookrightarrow (V_\pm,\partial V_\pm)$ (see [Wa2, pp. 39-40]), these handles being joined to the based disk of $V_0^\pm$ by an arc $\gamma_i^\pm$ in $\partial_\pm\bar{V}$. Hence, the elements e_i may be represented by disjoint 3-handles in V, these handles being the boundary connected sum of Σ_i^- with Σ_i^+ along the arc $\gamma_i^+ \cdot (\gamma_i^-)^{-1}$ (recall that r sends the based disk of V_0^+ to those of V_0^- diffeomorphically). As in [Wa2, pp. 40-41], one checks that, by deleting these handles out of $(\overline{V},\partial\overline{V})$, one obtains a homotopy equivalence $\bar{h} : (\overline{V},\partial\overline{V}) \longrightarrow (R,\partial R)$ which, by construction, is Poincaré cobordant to h.

We now turn our attention to the morphism of Q-spaces $f = g \circ h : V \longrightarrow P$. Write $f_0 = g \circ h_0 : V_0^- \longrightarrow P_0$. The map f_0 already induces isomorphisms $\pi_1(\partial V_0^-) \xrightarrow{\simeq} \pi_1(\partial P_0)$ and $\pi_1(V_0^-) \xrightarrow{\simeq} \pi_1(P_0)$.

By deleting 2- handles on $(V_0^-,\partial V_0^-)$, one obtains a normal map of degree one $\hat{f}_0 : \hat{V}_0^- \longrightarrow P_0$ with the same properties as f_0 with in addition $K_j(\hat{V}_0^-,\partial\hat{V}_0^-) = 0$ for $j \leq 2$. The maps f_0 and $\hat{f}_0$ are normally cobordant, namely there is a manifold cobordism $(W_0^-,V_0^-,\hat{V}_0^-)$ and a normal map of degree one $F_0 : W_0^- \longrightarrow R_0 \times [0,1]$ extending f_0 and $\hat{f}_0$. By surgeries on W_0^- and on $W_0^- - (V_0^- \cup \hat{V}_0^-)$, relative to V_0^-, one can assume that the homomorphisms $\pi_1(\partial W_0^- - (V_0^- \cup \hat{V}_0^-)) \longrightarrow \pi_1(W_0^-)$ and $\pi_1(\partial\hat{V}_0^-) \longrightarrow \pi_1(\hat{V}_0^-)$ are isomorphisms.

The lifting $j^{P_+} \circ f_0$, which gave rise to to the homotopy equivalence r extends to W_0^-. Therefore, there exists a normal map of degree one

$$\mu \ : \ (W_0^+, V_0^+, \hat{V}_0^+, \partial W_0^+ - (V_0^+ \cup \hat{V}_0^+)) \longrightarrow (W_0^-, V_0^-, \hat{V}_0^-, \partial W_0^- - (V_0^- \cup \hat{V}_0^-))$$

such that $\mu | V_0^+ = r$. Let $\hat{r} = \mu | \hat{V}_0^+$.

As $\pi_1(\partial\hat{V}_0^-) = \pi_1(\hat{V}_0^-)$, one can proceed with $\hat{r}$ as with the first map r : the map $\hat{r}$ is normally cobordant to a homotopy equivalence after addition to $\hat{V}_0^+$ and $\hat{V}_0^-$ of a set of trivial 2-handles. Therefore, by changing the cobordisms W_0^+ and W_0^-, one can assume that $\hat{r}$ is a homotopy equivalence.

By construction, the homomorphisms $\pi_1(\partial W_0^- - (V_0^- \cup \hat{V}_0^-)) \rightarrow \pi_1(W_0^-)$ is an isomorphism. As $\dim W_0^- = 6$, the classical $(\pi-\pi)$-theorem asserts that μ is normally cobordant (relative to $V_0^- \cup \hat{V}_0^-$) to a homotopy equivalence $\bar{\mu} : \bar{W}_0^+ \longrightarrow \bar{W}_0^-$. Let us construct the Poincaré cobordism W from V to $\hat{V}$ as follows :

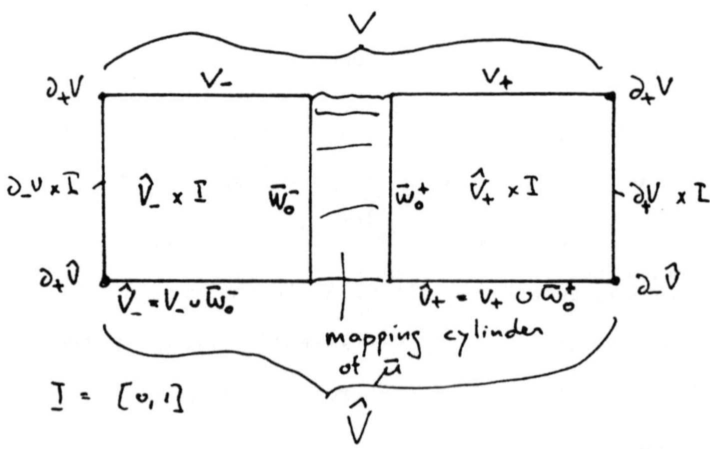

There is an extension $F : W \longrightarrow P$ of f, leading to $\hat{f} : \hat{V} \longrightarrow P$. The morphism of Q-spaces $\hat{f}$ has the same properties as $h : V \rightarrow R$, namely $\hat{V}$ is the union of two manifolds $\hat{V} = \hat{V}_- \cup \hat{V}_+$, $\hat{V}_- \cap \hat{V}_+ = \hat{V}_0$, $K_1(\hat{V}_0) = 0$ and $K_j(\hat{V}_0, \partial\hat{V}_0) = 0$ for $j \leq 2$. We can now proceed with $\hat{f}$ as we did with the map h. This enables us to to transform f into a homotopy equivalence, up to Poincaré cobordism. This shows that $f = g \circ h$ is Poincaré cobordant to id_P. As h was shown to be Poincaré cobordant to id_R, we have finally established that $g : R \longrightarrow P$ is Poincaré cobordant to id_P.

<u>Case n = 8</u> : The principle is the same as in the case n = 6; the technique is slightly simpler. By surgeries on ∂V_0^- and V_0^- and by handle substractions on the pair $(R_0, \partial R_0)$, one can assume that $\pi_1(\partial V_0^-) = \pi_1(V_0^-)$, $K_j(V_0^-) = 0$ for $j \leq 2$ and $K_j(V_0^-, \partial V_0^-) = 0$ for $j \leq 3$.

As $\dim V_0^- = 7$, the $(\pi\text{-}\pi)$-theorem holds (this is the simplification with respect to the case n = 6), and the bordism class of normal maps of degree one corresponding to the lifting $\tilde{j}^{R_-}|R_0$ contains a homotopy equivalence $r : V_0^+ \rightarrow V_0^-$, preserving some based disks. As in the case n = 6, form the Poincaré space $V = V_- \cup_r V_+$, where $V_\pm$ are manifolds with $\partial V_\pm = \partial_\pm V \cup V_0$. The map $h_0 : V_0^- \rightarrow R_0$ extends to a morphism of Poincaré space $h : V \rightarrow R$.

As $V_\pm$ are manifolds, one can perform surgeries (away of V_0) so that one obtains a morphism of Poincaré space $\bar{h} : \bar{V} = \bar{V}_- \cup \bar{V}_+ \to R$ such that

a) $\bar{h}$ induces an isomorphism on the fundamental groups of all the spaces under consideration ($\bar{V}_\pm, \partial_\pm\bar{V}$, V_0 and ∂V_0) (in fact $\bar{h}|V_0 = h_0$).

b) $K_j(\bar{V}_\pm) = 0 = K_j(\bar{V}_\pm, \partial_\pm\bar{V})$ for $j \leq 3$.

This enables us to proceed as in the case $n = 6$, namely deleting 4-handles out of $(\bar{V}, \partial\bar{V})$ to obtain a homotopy equivalence $\bar{\bar{h}} : (\bar{\bar{V}}, \partial\bar{\bar{V}}) \longrightarrow (R, \partial R)$ which is Poincaré cobordant to h.

We now continue as in the case $n = 6$, to prove that $f = g \circ h$ is Poincaré cobordant to id_P. Again, the argument is simpler since the map $\hat{f} : \hat{V}_0^+ \longrightarrow \hat{V}_0^-$ is normally cobordant to a homotopy equivalence by the $(\pi\text{-}\pi)$-theorem (we do not have to introduce the additional set of handles). Details are left to the reader.

<u>Case $n = 7$</u> : By surgeries on ∂V_0^- and V_0^- and by handle substractions on the pair $(R_0, \partial R_0)$, one can assume that $\pi_1(\partial V_0^-) \cong \pi_1(V_0^-)$, $K_1(V_0^-) \equiv 0$ and $K_j(V_0^-, \partial V_0^-) = 0$ for $j \leq 3$ (by deleting 3-handles, which is possible since $\pi_1(\partial V_0^-) = \pi_1(V_0^-)$; observe that this operation may create elements in $K_2(V_0^-)$ and for this reason we only assume that $K_1(V_0^-) = 0$.

As in the case $n = 8$, one can use the $(\pi\text{-}\pi)$-theorem (since $\dim V_0^- = 6$) and then the bordism class of normal maps of degree one corresponding to the lifting $\hat{j}^{R}|R_0$ contains a homotopy equivalence $r : V_0^+ \longrightarrow V_0^-$, preserving some based disks. As in the cases $n = 6$ or 8, form the Poincaré space $V = V_- \cup_r V_+$, where $V_\pm$ are manifolds with $\partial V_\pm = \partial_\pm V \cup V_0$. The map $h_0 : V_0^- \rightarrow R_0$ extends to a morphism of Poincaré space $h : V \longrightarrow R$.

As $V_\pm$ are manifolds, one can perform surgeries (away of V_0) so that one obtains a morphism of Poincaré space $\bar{h} : \bar{V} = \bar{V}_- \cup \bar{V}_+$ R such that

a) $\bar{h}$ induces an isomorphism on the fundamental groups of all the spaces under consideration ($\bar{V}_\pm, \partial_\pm \bar{V}$, V_0 and ∂V_0) (in fact $\bar{h}|V_0 = h_0$).

b) $K_j(\bar{V}_\pm) = 0$ for $j \leq 2$ and $K_j(\bar{V}_\pm, \partial_\pm \bar{V})$ for $j \leq 3$.

The Mayer-Vietoris sequences of (4.16) show that $K_j(\bar{V}) = 0$ for $j \leq 2$ and that $K_j(\bar{V}, \partial\bar{V}) = 0$ for $j \leq 3$. Using Poincaré duality (see the case $n = 6$), one shows that the only kernels of $\bar{h}$ which might not vanish are those of the following exact sequence :

$$0 \longrightarrow K_4(\bar{V}, \partial\bar{V}) \longrightarrow K_3(\partial\bar{V}) \longrightarrow K_3(\bar{V}) \longrightarrow 0$$

and all of them are stably $Z\pi$-free. By replacing $\bar{V}$ by its boundary connected sum with copies of $D^4 \times S^3$, one may assume that $K_4(\bar{V}, \partial\bar{V})$ is $Z\pi$-free.

Since the pairs $(\overline{V}_{\pm},\partial_{\pm}\overline{V})$ are 2-connected, it is possible to represent any element of $\pi_4(\overline{V}_{\pm},\partial_{\pm}\overline{V})$ by a 4-handle of $(\overline{V}_{\pm},\partial_{\pm}\overline{V})$: use [Hn, Theorem 8.1] to PL-embedd a 4-disk and smooth this embedding. As in the previous cases ($n = 6$ or 8), this enables us to realize a basis of $K_4(\overline{V},\partial\overline{V})$ by a set of disjoint 4-handles of $(\overline{V}_{\pm},\partial_{\pm}\overline{V})$. Using Poincaré duality, one checks that deleting these handles gives rise to a homotopy equivalence $\bar{\bar{h}} : (\bar{\bar{V}},\partial\bar{\bar{V}}) \longrightarrow (R,\partial R)$ which is Poincaré cobordant to h.

Again, similarly to the case $n = 8$, we reapply our argument to the map $f = g \circ h$ to prove that this map is Poincaré cobordant to id_P. This shows that g is Poincaré cobordant to id_P.

This terminates the proof of Lemma (4.17). [ ]

After Lemmas (4.15) and (4.17) (using (4.11.b) and (4.14)), one has

(4.18) <u>**Lemma**</u> **RA**(0,9) is true.

e) THE ASSERTION S(i,n)

In order to propagate Asertion $\Pi(i,n)$ along with **RA**(i,n), we introduce Assertion **S**(i,n) below. This assertion provides us with the ability to perform surgeries and handle substractions of order to prove $\Pi(i,n)$, by an argument similar to those of Lemma (4.17).

(4.19) **Assertion** S(i,n) : Let (X,Y) be a pair over BG such that j^X admits a lifting through B_i. Let P be a Poincaré space of formal dimension n with a Poincaré decomposition of its boundary : $\partial P = L_A \cup L_B$. Suppose that the pairs (P,L_B) and $(L_B,\partial L_B)$ are 2-connected. Let $f : (P,L_B) \longrightarrow (X,Y)$ be a map over BG. Then, the homotopy kernels $\widetilde{K}_j(P)$ and $\widetilde{K}_j(P,L_B)$ of f satisfy the following properties :

a) Let $a \in \widetilde{K}_j(P)$, with $j \leq [\frac{n}{2}]-1$. Then a Poincaré surgery can be performed on a.

b) Let $e \in \widetilde{K}_j(P,L_B)$ with $j \leq [\frac{n+1}{2}]$. Then a Poincaré handle subtraction can be performed on e, relative to L_A.

We denote by $S_\emptyset(i,n)$ the special case of S(i,n) when $L_A = \emptyset$.

Assertion S(i,n) will be proven in a linked way with Assertion **RA**(i,n) by a double induction on i and n. In addition to their role in proving **RA**(i,n), Assertions S(i,n) and $S_\emptyset(i,n)$ will also be used in Chapter 5. The induction on n for S(i,n) starts with a special argument in dimensions ≤ 8 :

(4.20) **Lemma** $S_\emptyset(i,n)$ is true for all i, when n = 6, 7 or 8.

Proof : The proof is independent of i. Let $f : P \longrightarrow X$ be a map as in the statement of $S_\emptyset(i,n)$. Since the pair $(P,\partial P)$ is 2-connected and $n \leq 8$, there is a lifting :

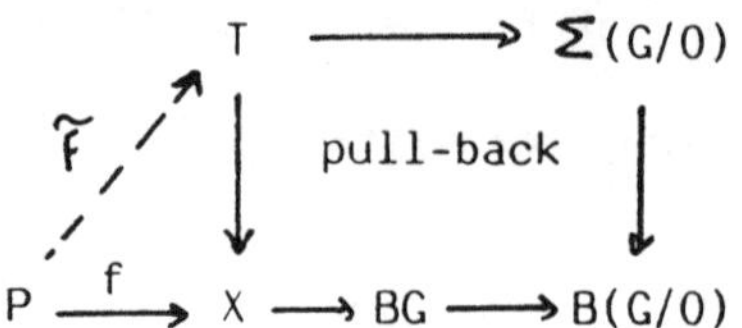

(see the proof of Lemma (4.17)). The decomposition of $\Sigma(G/0)$ into two cones gives rise to a decomposition $T = T_- \cup T_+$ and a Q-decomposition

$$P = P_- \cup P_+, \quad P_- \cap P_+ = P_0, \quad \partial P_\pm = \partial_\pm P \cup P_0$$

respected by f, and $j^T{\pm}$ admits a lifting $\tilde{j}^T{\pm} : T_\pm \longrightarrow BO$. Using Lemma (4.17) and its proof, one obtains a homotopy equivalence $h : V \longrightarrow P$ (this is the map called $\bar{h}$ in the proof of (4.17)), such that :

a) V is a Poincaré space obtained by gluing two manifolds V_- and V_+ $(\partial V_\pm = \partial_\pm V \cup V_0^\pm)$ by a homotopy equivalence $V_0^+ \longrightarrow V_0^-$; one has $h(V_\pm) \subset P_\pm$

b) The homotopy kernels of h satisfy :

$$\tilde{K}_j(V_\pm) = 0 \qquad \text{for} \quad j \leq [\tfrac{n}{2}]-2$$

$$\tilde{K}_j(V_\pm,\partial_\pm V) = 0 \qquad " \qquad j \leq [\tfrac{n-1}{2}].$$

Let $a \in \widetilde{K}_j(P)$ represented by a commutative diagram

$$\begin{array}{ccc} S^j & \xrightarrow{\alpha} & P \\ \cap & & \downarrow f \\ D^{j+1} & \xrightarrow{\bar{\alpha}} & X \end{array}$$

One can see $\bar{\alpha}$ as a homotopy of $f \circ \alpha$ to a constant map to a point in the image of, say, T_-. The homotopy lifting property applied twice enables us to show that a may be represented by a commutative diagram

$$\begin{array}{ccc} S^j & \xrightarrow{\alpha} & P_- \\ \cap & & \downarrow \bar{f} \\ D^{j+1} & \xrightarrow{\bar{\alpha}} & T_- \end{array}$$

If $j \leq [\frac{n}{2}]-1$, one has $\pi_j(h|V_-) = K_{j-1}(V_-) = 0$. Therefore, there exists a map $\beta : S^j \to V_-$ such that $h \circ \beta$ is homotopic to α_-. As V_- is a manifold, the map β is homotopic to an embedding and therefore the class a can be realized by some PD-embedding. By Proposition (3.14), a Poincaré surgery may be performed on a, which proves Part a).

Part b) of Assertions $S_\emptyset(i,n)$ $(n = 6, 7 \text{ or } 8)$ is proven by a completely similar argument. Details are left to the reader. []

(4.21) **Lemma** **S**(i,7) and **S**(i,8) are true for all i.

Proof : The principle is the same as for Lemma (4.20). Let $f : P \to X$ be a map as in the statement of **S**(i,n). Since the pair $(P,\partial P)$ is 2-connected and $n \leq 8$, there is a lifting :

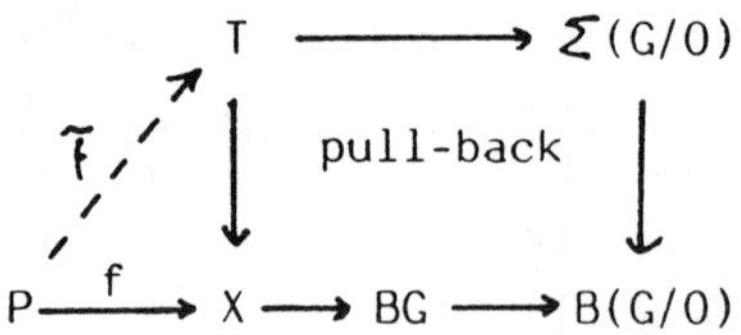

(see the proof of Lemma (4.17)). The decomposition of $\Sigma(G/0)$ into two cones gives rise to a Q-decompositions :

$$P = P_- \cup P_+, \qquad P_- \cap P_+ = P_0, \qquad \partial P_\pm = \partial_\pm P \cup P_0,$$

$$L_A = L_A^- \cup L_A^+, \qquad L_A^- \cap L_A^+ = L_A \cap P_0, \quad L_A^\pm = L_A \cap P_\pm,$$

$$L_B = L_B^- \cup L_B^+, \qquad L_B^- \cap L_B^+ = L_B \cap P_0, \quad L_B^\pm = L_B \cap P_\pm,$$

and $j^{P_\pm}$ admits a lifting $j^{P_\pm} : P_\pm \longrightarrow BO$.

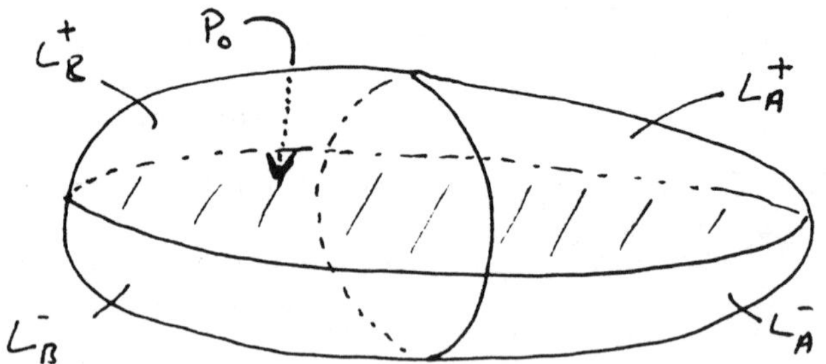

The proof of Lemma (4.20) applied to id_{L_A} shows that id_{L_A} is Poincaré cobordant to a homotopy equivalence $\gamma : V_A \longrightarrow L_A$, where the Poincaré space V_A is obtained by gluing two manifolds V_A^- and V_A^+ $(\partial V_A^\pm = \partial_\pm V_A \cup V_{A,0}^\pm)$ by a homotopy equivalence $r_A : V^+_{A,0} \rightarrow V^-_{A,0}$ and one has $\gamma(V_A^\pm) = L_A^\pm$. Using γ, one can change P up to homotopy equivalence and consider that $L_A = V_A$.

Using the same argument as in the proof of (4.20), one can change by a homotopy equivalence (relative to V_A) the Q-decompositions of (P,L_B) into a Poincaré decomposition in which the pieces $P_\pm$ and $L_B^\pm$ are manifolds. The same construction as for proving $S_\emptyset(i,n)$ in Lemma (4.20) can now be performed (relative to V_A) to establish the conclusions of $S(i,n)$ for n = 7 or 8. As the arguments are completely similar, details are left to the reader. []

(4.22) **Lemma** $S(0,n)$ is true for all $n \geq 7$

Proof : We first establish $S_\emptyset(0,n)$. Let $f : P \longrightarrow X$ be a map as in the statement of Assertion $S_\emptyset(0,n)$. Since $(P,\partial P)$ is 2-connected and since j^P admits a lifting through BO, the classical $(\pi$-$\pi)$-theorem $(n \geq 6)$ asserts that P is homotopy equivalent to a manifold. The conclusions of $S_\emptyset(0,n)$ then follows from the classical embedding theory and from (3.14) or (3.17).

The proof of $S(0,n)$ goes in the same way. We need here $n \geq 7$, since we use first the $(\pi$-$\pi)$-theorem in dimension n-1 to replace $L_A \subset \partial P$ by a manifold up to homotopy equivalence. []

f) THE FINAL INDUCTION LEMMA

We introduce a assertion :

Assertion **RAS**(i,n) : **RA**(i,n) is true and S(j,k) is true for $(0,7) \preccurlyeq (j,k) \preccurlyeq (i,n)$.

Observe that if $(i,n) \prec (0,7)$, Assertion **RAS**(i,n) coincides with **RA**(i,n).

(4.23) **Lemma** RAS(i-1,n) implies **RAS**(i,n).

Before proving Lemma (4.23), we draw its consequences on our various assertions.

(4.24) **Proposition**

a) **A**(i,n) is true for all i and n.

b) π(i,n) is true for all i and n.

c) $S_\emptyset(i,n)$ is true for $(i,n) \succcurlyeq (0,6)$

d) S(i,n) is true for $(i,n) \succcurlyeq (0,7)$.

Proof : By Lemmas (4.18), (4.21) and (4.22), one has that **RAS**(0,9) is true. Hence, Lemmas (4.22) and (4.23) imply that **RAS**(i,n) is true for all (i,n). This proves Part a) and d) of (4.24). Part c) then follows from (4.20). Part b) is a consequence of Part a) (see See subsection c) above). []

(4.25) **Remarks** : a) The fact that $\pi(i,n)$ is weaker as **A**(i,n) is the reason of the non-appearance of $\pi(i,n)$ in Assertion **RAS**(i,n) and in Lemma (4.23). But in fact, in the proof of (4.23), we first establish that **RAS**$(i-1,n)$ implies $\pi(i,n)$ and then use Lemma (4.15) to obtain **A**(i,n). Observe that Assertion $\pi(i,n)$ is not interesting in itself, since it will eventually be covered by Proposition (5.2).

b) In consequence of Part a), Proposition (4.8) is proven (see subsection a)).

Proof of Lemma (4.23) : Since **RAS**$(0,9)$ is true, one may assume that $n \geqslant 9$. By Lemmas (4.11) and (4.22), it suffices to give an argument for $i \geqslant 1$. The proof divides into two steps.

Step 1 : **RAS**$(i-1,n)$ implies **RA**(i,n).

By Lemma (4.15), it is enough to prove that **RAS**$(i-1,n)$ implies $\pi(i,n)$. Let $g : (R,\partial R) \longrightarrow (P,\partial P)$ be a morphism of Poincaré spaces as in the statement of $\pi(i,n)$. By low-dimensional Poincaré surgeries (Lemmas (3.16) or (3.19)) one may assume that $(R,\partial R)$ is 2-connected.

As in the proofs of (4.12) and (4.17), one obtains Q-decompositions :

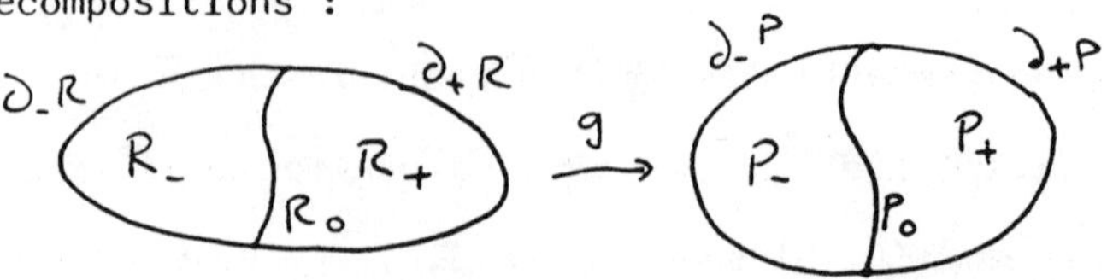

so that j^P_- and j^P_+ admits liftings $\tilde{j}^P_-$ and $\tilde{j}^P_+$ through B_{i-1}. Moreover, the pairs $(R_0,\partial R_0)$, $(R_\pm,\partial_\pm R)$, $(P_0,\partial P_0)$ and $(P_\pm,\partial_\pm P)$ are 2-connected. All the spaces under consideration are homotopy equivalent to complexes with finite skeleta by the results of the appendix, at the end of these notes.

As the pair $(R_0,\partial R_0)$ is 2-connected and, by **RA**(i-1,n), $\mathbf{A}^e$(i-1,n-1) holds true, the class $[id : R_0 \longrightarrow R_0]$ is in the image of $\Omega^P_{n-1}(R_0,\partial R_0) \longrightarrow \Omega^Q_{n-1}(R_0,\partial R_0)$. Therefore, there is a Q-cobordism $(W_0,R_0,\overline{R}_0)$ and a retraction $h_0 : W_0 \longrightarrow R_0$ such that $\overline{R}_0$ is a Poincaré space. Using $\mathbf{S}_\emptyset$(i-1,n-1), one can perform Poincaré surgeries on $\overline{R}_0$ and handle substractions on $(\overline{R}_0,\partial\overline{R}_0)$ so that :

$$K_j(\overline{R}_0) = 0 \text{ for } j \leq [\tfrac{n-2}{2}] - 1$$

$$K_j(\overline{R}_0,\partial\overline{R}_0) = 0 \text{ for } j \leq [\tfrac{n-1}{2}]$$

(Observe that it is not possible to obtain a better result in absence of Poincaré duality for R_0; indeed, deleting a handle of index $[\frac{n-1}{2}]$ on $(\overline{R}_0,\partial\overline{R}_0)$ will in general create elements in $K_j(\overline{R}_0)$ for $j = [\frac{n-2}{2}]$.)

The maps $h_{\pm} = id \cup h_0 : R_{\pm} \cup W_0 \longrightarrow R_{\pm}$ represent elements of $\Omega_n^{QP}(R_{\pm},\partial_{\pm}R)$. By **A**(i-1,n), the groups $\Omega_n^{QP}(R_{\pm},\partial_{\pm}R)$ vanish and there are Q-cobordisms relative to $\bar{R}_0$, from $h_{\pm}$ to $\bar{h}_{\pm}$: $(\bar{R}_{\pm},\partial_{\pm}\bar{R},\bar{R}_0) \longrightarrow (R_{\pm},\partial_{\pm}R,R_0)$ such that $\bar{R}_{\pm}$ are Poincaré spaces with boundary Poincaré décomposed : $\partial\bar{R}_{\pm} = \partial_{\pm}\bar{R} \cup \bar{R}_0$. The Poincaré space $\bar{R} = \bar{R}_- \cup \bar{R}_+$ is now Poincaré decomposed and the maps $\bar{h}_{\pm}$ give rise to a morphism of Poincaré spaces $\bar{h} : \bar{R} \longrightarrow R$ respecting the decompositions. Using **S**(i-1,n), it is possible to change $\bar{R}$ up to Poincaré cobordism (relative to $\bar{R}_0$) so that the new map $\bar{h} : \bar{R} \longrightarrow R$ satisfies :

$$\widetilde{K}_j(\bar{R}_{\pm}) = 0 \text{ for } j \leq [\tfrac{n}{2}] - 1$$

$$\widetilde{K}_j(\bar{R}_{\pm},\partial\bar{R}_{\pm}) = 0 \text{ for } j \leq [\tfrac{n-1}{2}]$$

Suppose now that $n = 2k$. The Mayer-Vietoris sequences (4.16) and Poincaré duality on R imply that the only homology kernels of $\bar{h}$ which might not vanish are those of the following exact sequence :

$$0 \longrightarrow K_k(\partial\bar{R}) \longrightarrow K_k(\bar{R}) \longrightarrow K_k(\bar{R},\partial\bar{R}) \longrightarrow K_{k-1}(\partial\bar{R}) \longrightarrow 0$$

Moreover, $K_k(\bar{R})$ and $K_k(\bar{R},\partial\bar{R})$ are stably $Z\pi$-free. Make $K_k(\bar{R},\partial\bar{R})$ $Z\pi$-free, with basis, say : $e_1,..,e_r$, by taking the boundary connected sum of $\bar{R}$ with copies of $D^k \times S^k$. By the Mayer-Vietoris sequences, the homomorphism

$$K_k(\bar{R}_-,\partial_-\bar{R}) \oplus K_k(\bar{R}_+,\partial_+\bar{R}) \longrightarrow K_k(\bar{R},\partial\bar{R})$$

is surjective. Let $(e_t^-,e_t^+) \in K_k(\bar{R}_-,\partial_-\bar{R}) \oplus K_k(\bar{R}_+,\partial_+\bar{R})$ having image e_t in $K_k(\bar{R},\partial\bar{R})$ ($t = 1$ to r). Since $K_j(R_\pm,\partial_\pm R) = 0$ for $j \leq k-1$, one has $K_k(\bar{R}_\pm,\partial_\pm\bar{R}) = \pi_{k+1}(h|(\bar{R}_\pm,\partial_\pm\bar{R})$. Using $S(i-1,n)$, one can realize each $e_t^\pm$ (one after another) by a Poincaré handle of $(\bar{R}_\pm,\partial_\pm\bar{R})$ (relative to $\bar{R}_0$) and thus get a set of disjoint Poincaré k-handles of $(\bar{R}_\pm,\partial_\pm\bar{R})$. Each of these handles is attached by a path $\gamma_t^\pm$ in $\partial_\pm\bar{R}$ to a based disk of $\partial\bar{R}_0$. By (3.19), the paths $\gamma_t^+\cdot(\gamma_t^-)^{-1}$ can be realized by a 1-handle in $\partial\bar{R}$ (and the handles corresponding to different t's are disjoint). Therefore, the set $\{e_t \mid t = 1,\ldots,r\}$ can be realized by a set of disjoint k-handles in $(\bar{R},\partial\bar{R})$, the k-handle corresponding to e_t being the connected sum of the handles corresponding to e_t^- and e_t^+, along the 1-handle realizing $\gamma_t^+\cdot(\gamma_t^-)^{-1}$ (up to homotopy type, one may assume that $\partial\bar{R}$ has a collar neighbourhood in $\bar{R}$ and thus this connected sum is done in the same way as in a manifold). Deleting these handles produces a homotopy equivalence $\bar{\bar{h}} : (\bar{\bar{R}},\partial\bar{\bar{R}}) \longrightarrow (R,\partial R)$ which is Poincaré bordant to h (as in the proof of (4.17), we follow the argument of [Wa2, pp. 40,41]).

In the case $n = 2k+1$, we are left with the kernels

$$0 \longrightarrow K_{k+1}(\bar{R}) \longrightarrow K_{k-1}(\bar{R},\partial\bar{R}) \longrightarrow K_k(\partial\bar{R}) \longrightarrow K_k(\bar{R}) \longrightarrow 0$$

and $K_{k+1}(\bar{R},\partial\bar{R})$ is stably $Z\pi$-free. We can then proceed as in the case $n = 2k$, but with $(k+1)$-handles of $(\bar{R}_{\pm},\partial_{\pm}\bar{R})$. Details are completely similar, following the procedure of the proof of Lemma (4.7), case $n = 7$.

In both cases, we have constructed a homotopy equivalence $h : (\bar{\bar{R}},\partial\bar{\bar{R}}) \longrightarrow (R,\partial R)$ where R is Poincaré decomposed $\bar{\bar{R}} = \bar{\bar{R}}_- \quad \bar{\bar{R}}_+$, etc, and the decompositions are respected by h.

We can now play with $f = g \circ h : (\bar{\bar{R}},\bar{\bar{R}}_{\pm},\partial_{\pm}\bar{\bar{R}},\bar{\bar{R}}_0) \longrightarrow (P,P_{\pm},\partial_{\pm}P,P_0)$ the same game as was played with h (details as in the proof of (4.17), case $n = 7$ or 8). This will provide a homotopy equivalence $\hat{f} : \hat{R} \longrightarrow P$ which is Poincaré cobordant to f. As $\hat{f}$ is Poincaré cobordant to id_P and h is Poincaré cobordant to id_R (they are both homotopy equivalences), one deduces that g is Poincaré cobordant to id_P, which proves $\Pi(i,n)$.

Step 2 : **RAS**$(i-1,n)$ imples $S(i,n)$.

We first give the simpler argument for $\mathbf{S}_{\emptyset}(i,n)$. Let $f: (P,L_B) \rightarrow (X,Y)$ be as in the statement of $\mathbf{S}_{\emptyset}(i,n)$. We apply the proof of Step 1 to $R = P$ and $g = id$, using decompositions $P_{\pm} \longrightarrow X_{\pm}$ obtained as in the proof of (4.12). Observe that the homotopy kernels of the map $\bar{\bar{h}} : \bar{\bar{P}} \longrightarrow P$ $(P = R)$satisfy :

$$\tilde{K}_j(\bar{\bar{P}}_\pm) = 0 \qquad \text{for} \quad j \leq [\tfrac{n}{2}] - 2$$

$$\tilde{K}_j(\bar{P}_\pm, \partial_\pm \bar{P}) = 0 \qquad \text{"} \quad j \leq [\tfrac{n-1}{2}]$$

Let $a \in \tilde{K}_j(P)$ be represented by a commutative diagram

$$\begin{array}{ccc} S^j & \xrightarrow{\alpha} & P \\ \cap & & \downarrow f \\ D^{j+1} & \xrightarrow{\bar{\alpha}} & X \end{array}$$

One can see $\bar{\alpha}$ as a homotopy of $f \circ \alpha$ to a constant map to a point belonging, say, to X_-). The homotopy lifting property enables us to show that a may be represented by a commutative diagram

$$\begin{array}{ccc} S^j & \xrightarrow{\alpha_-} & P_- \\ \cap & & \downarrow f|P_- \\ D^{j+1} & \xrightarrow{\bar{\alpha}_-} & X_- \end{array}$$

If $j \leq [\frac{n}{2}]-1$, one has $\pi_j(\bar{\bar{h}}|\bar{P}_-) = K_{j-1}(\bar{\bar{P}}_-) = 0$. Therefore, there exists a map $\beta : S^j \longrightarrow \bar{\bar{P}}_-$ such that $h \circ \beta$ is homotopic to α_- and then a is the image of some element $\bar{a} \in \pi_{j+1}(f \circ \bar{\bar{h}}|\bar{\bar{P}}_- : \bar{P}_- \longrightarrow X_-)$. As j^{X_-} admits a lifting through B_{i-1} and $\mathbf{S}(i-1,n)$ holds true, a Poincaré surgery may be performed on $\bar{a}$ and then on a. This proves Part a) of $\mathbf{S}_\emptyset(i,n)$.

Part b) of Assertion $\mathbf{S}_\emptyset(i,n)$ is proven by a completely similar argument. Details are left to the reader.

The case $L_A \neq \emptyset$ is also similar. Let $f : P \longrightarrow X$ be a map as in the statement of **S**(i,n), with $L_A \neq \emptyset$. As in Step 2, we apply the proof of Step 1 to $R = P$ and $g = id$. But, in addition to the Q-decomposition $P = P_- \cup P_+$, we obtain induced Q-decompositions :

$$L_A = L_A^- \cup L_A^+ , \quad L_A^- \cap L_A^+ = L_A \cap P_0, \quad L_A^{\pm} = L_A \cap P_{\pm} ,$$

$$L_B = L_B^- \cup L_B^+ , \quad L_B^- \cap L_B^+ = L_B \cap P_0, \quad L_B^{\pm} = L_B \cap P_{\pm} .$$

and $j^P | L_A^{\pm}$ admits a lifting through B_{i-1}. Using **A**(i-1,n-2), **A**(i-1,n-1), $\mathbf{S}_{\emptyset}$(i-1,n-2) and **S**(i-1,n-1), one proves as in Step 1 that L_A is homotopy equivalent to $\overline{L}_A$ which is Poincaré decomposed : $\overline{L}_A = \overline{L}_A^- \cup \overline{L}_A^+$, etc. This decomposition is preserved by the homotopy equivalence $\overline{L}_A \longrightarrow L_A$. Therefore, it is possible to replace L_A by $\overline{L}_A$ and assume that the Q-decomposition of L_A is actually a Poincaré decomposition.

As L_A is already Poincaré decomposed, it is possible to prove the conclusions of **S**(i,n) by using the same argument as in Step 2, but relative to L_A. []

5. SURGERY ON POINCARE SPACES

A. THE FORMAL POINT OF VIEW

Let P be a Poincaré space of formal dimension n with a Poincaré decomposition $\partial P = \partial_- P \cup \partial_+ P$ of ∂P. Let $\mathcal{N}(P,\partial_+ P)$ be the set of Poincaré bordism classes of morphisms of Poincaré spaces :

$$f : (\bar{P},\partial_-\bar{P},\partial_+\bar{P}) \longrightarrow (P,\partial_- P,\partial_+ P)$$

where $\partial\bar{P} = \partial_-\bar{P} \cup \partial_+\bar{P}$ is a Poincaré decomposition of $\partial\bar{P}$ and $f| \partial_-\bar{P} : \partial_-\bar{P} \longrightarrow \partial_- P$ is a homotopy equivalence. Observe that, if j^P admits a lifting $\tilde{j}^P$ through BO, if $\bar{P}$ is a manifold, and if $\tilde{j}^P \circ f$ classifies the normal bundle of $\bar{P}$, then f is a normal map of degree one in the sense of the classical surgery theory. We then say that f is a **classical surgery problem**. The main result of this section is the following theorem :

(5.1) **Theorem** There exists a map :

$$s : \mathcal{N}(P,\partial_+ P) \longrightarrow L_n(\pi(P),\pi(\partial_+ P);w^P)$$

such that :

1) Suppose that $\partial_+ P = \emptyset$ or $n \geqslant 4$. Then, $s(\alpha) = 0$ if and only if α contains a representative $f : \bar{P} \longrightarrow P$ such that f and $f|\partial_+ P$ are homotopy equivalences.

2) Suppose that f preserves Poincaré decompositions :

$\overline{P} = \overline{P}_1 \cup \overline{P}_2$, $\overline{P}_1 \cap \overline{P}_2 = \overline{P}_0$, $\partial\overline{P}_i = \partial_i\overline{P} \cup \overline{P}_0$,

$\partial_- P \subset \mathrm{int}\partial_1 P$,

$P = P_1 \cup P_2$, $P_1 \cap P_2 = P_0$, $\partial P_i = \partial_i P \cup P_0$,

$\partial_- P \subset \mathrm{int}\partial_1 P$,

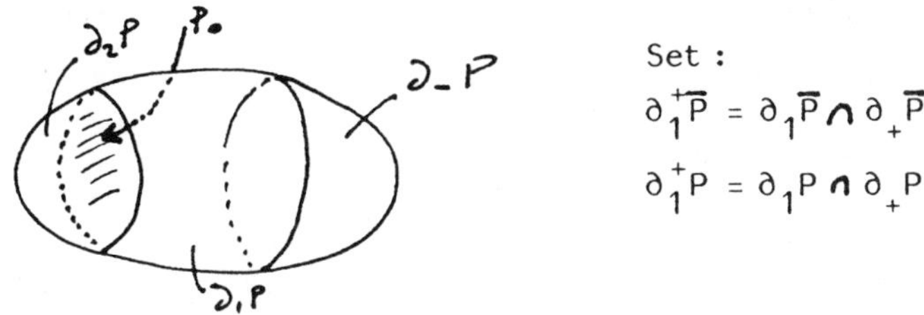

Set :

$\partial_1^+\overline{P} = \partial_1\overline{P} \cap \partial_+\overline{P}$

$\partial_1^+ P = \partial_1 P \cap \partial_+ P$

Denote by $f_i : \overline{P}_i \longrightarrow P_i$ the restriction $f|P_i$. Suppose that $f_0 : (\overline{P}_0, \partial\overline{P}_0) \longrightarrow (P_0, \partial P_0)$ is a homotopy equivalence. Denote by :

$$t_1 : L_n(\pi(P_1), \pi(\partial_1^+ P); w^{P_1}) \longrightarrow L_n(\pi(P), \pi(\partial_+ P); w^P)$$

$$t_2 : L_n(\pi(P_2), \pi(\partial_2 P); w^{P_2}) \longrightarrow L_n(\pi(P), \pi(\partial_+ P); w^P)$$

the homomorphisms induced by the inclusions. Then :

$$s(f) = t_1(s(f_1)) + t_2(s(f_2)).$$

3) In the case of a classical surgery problem, s(f) coincides with the Wall surgery obstruction $\sigma(f)$.

Observe that the dimensional restrictions in Conclusion 1) of (5.1) are weaker than the ones appearing in classical surgery theory. In particular, there is no low-dimensional restriction when $\partial_+ P = \emptyset$.

As $L_n(\pi_1(P),\pi_1(\partial_+ P);w^P) = 0$ if $\pi_1(\partial_+ P) \cong \pi_1(P)$, (say P connected), Theorem (5.1) implies the following

(π-π)-theorem :

(5.2) **Corollary** Let P be a connected Poincaré space of formal dimension $n \geq 4$. Suppose that $\partial_+ P$ is connected and that $\pi_1(\partial_+ P) \cong \pi_1(P)$. Then any class in $\mathcal{N}(P,\partial_+ P)$ contains a homotopy equivalence.

Proof of (5.1) : Let $f : \bar{P} \longrightarrow P$ represent a class $[f] \in$ $(P,\partial_+ P)$. The map f extends to $F : \mathcal{M}_f \longrightarrow P$, where $\mathcal{M}_f$ is the mapping cylinder of f. By (3.8) and its proof, $\mathcal{M}_f$ is a Q-space of formal dimension n+1, with $\partial \mathcal{M}_f =$ $= \bar{P} \cup \mathcal{M}_{f|\partial \bar{P}} \cup P$. There is a Q-decomposition $\partial \mathcal{M}_f = \partial_- \mathcal{M}_f \cup$ $\partial_+ \mathcal{M}_f$, where :

$$\partial_- \mathcal{M}_f = \bar{P} \cup \mathcal{M}_{f|\partial_- \bar{P}} \cup P \quad (\partial_- \mathcal{M}_f \text{ is a Poincaré space})$$

$$\partial_+ \mathcal{M}_f = \mathcal{M}_{f|\partial_+ P}.$$

Therefore, $(\mathcal{M}_f,F)$ represents an element of $\Omega^{QP}_{n+1}(P,\partial_+ P)$. Define s(f) as :

$$s(f) = \sigma_n(\mathcal{M}_f,F)$$

where $\sigma_n : \Omega^{QP}_{n+1}(P,\partial_+ P) \longrightarrow L_n(\pi(P),\pi(\partial P);w^P)$ is the homomorphism of Theorem (4.4). It is easely checked that s is well defined.

If $f : (\overline{P},\partial_+\overline{P}) \longrightarrow (P,\partial_+P)$ is a homotopy equivalence, then $\mathcal{M}_f$ and $\mathcal{M}_{f|\partial_+P}$ are Poincaré spaces and $(\mathcal{M}_f,F)$ represents zero in $\Omega^{QP}_{n+1}(P,\partial_+P)$. Conversely, if $s(f) = 0$, then $(\mathcal{M}_f,F)$ represents zero in $\Omega^{QP}_{n+1}(P,\partial_+P)$, since, by Theorem (4.4), the homomorphism σ_n is injective when $\partial_+P = \emptyset$ or $n \geqslant 4$. Therefore, there exists :

a) a Q-cobordism (W,V) from $(\mathcal{M}_f,\partial\mathcal{M}_f)$ to $(Z,\partial Z)$, where Z is a Poincaré space

b) a Q-decomposition : $V = V_- \cup V_+$, where V_- is a Poincaré cobordism from $\partial_-\mathcal{M}_f$ to a Poincaré space ∂_-Z and V_+ is a Q-cobordism from $\partial_+\mathcal{M}_f$ to a Poincaré space ∂_+Z ($\partial Z = \partial_-Z \cup \partial_+Z$ is a Poincaré decomposition).

c) a map $G : (W,V_-,V_-) \longrightarrow (P,\partial_-P,\partial_+P)$ extending F.

Hence, $(V_- \cup Z,G)$ constitutes a Poincaré cobordism from $(\overline{P},f)$ to (P,id_P), (relative to ∂_-P). This shows that $[f] = [id_P]$ in $\eta(P,\partial_+P)$ and proves Part 1).

To prove Part 2), denote by

$$\tau_1 \ : \ (P_1,\partial_1^-P) \subset (P,\partial_+P)$$
$$\tau_2 \ : \ (P_2,\partial_2P) \subset (P,\partial_+P)$$

the inclusions inducing t_1 and t_2. It is easily checked that $(\overline{P},f)$ is Poincaré cobordant to $(\overline{P}_1 \amalg \overline{P}_2,\tau_1\circ f_1 \cup \tau_2\circ f_2)$. Therefore :

$$s(f) = \sigma_n(\mathcal{M}_f, F) = \sigma_n(\mathcal{M}_{\tau_1 \circ f_1}, F_1) + \sigma_n(\mathcal{M}_{\tau_2 \circ f_2}, F_2) =$$
$$= t_1(\sigma_n(\mathcal{M}_{f_1}, F_1)) + t_2(\sigma_n(\mathcal{M}_{f_2}, F_2)) =$$
$$= t_1(s(f_1)) + t_2(s(f_2)).$$

(the third equality comes from the naturality of σ_n). Thus, Part 2) is established.

To prove Part 3), let us consider $f : M \longrightarrow P$ a classical surgery problem (we give here the details for $\partial_+ P = \emptyset$, to simplify the exposition). This determines a lifting $\tilde{j}^P$ of j^P through BO and therefore $\tilde{j}^P \circ F$ is a lifting of $j^{\mathcal{M}_f}$ through BO. The element $s(f) = \sigma_n(\mathcal{M}_f, F)$ can be computed as in the proof of (4.7), using (4.9) : by transversality, one finds a normal map $g : W \longrightarrow \mathcal{M}_f$ and $\sigma_n(\mathcal{M}_f, F) = F_*(\sigma(g|\partial W))$, where σ is the Wall obstruction. But here we can take $W = M \times [0,1]$ and g to be the projection $M \times [0,1] \longrightarrow \mathcal{M}_f$. But then, $g|M \times 0 \cup \partial M \times [0,1]$ is a homotopy equivalence and $g|M \times 1 = f$. Therefore, $s(f) = \sigma(f)$. []

It is interesting to know whether the Poincaré surgery obstruction of Theorem (5.1) coincides with the other ones previously defined in the literature (see, for instance, [Ho1] , [Ha2], [Jo1]). An answer is given by the following theorem :

(5.3) **Theorem** For $n \geq 6$, the map s of (5.1) is characterized by properties 1), 2) and 3) of (5.1).

The proof of (5.3), as well as the arguments in the next chapters, make use of the following proposition :

(5.4) **Proposition** Let Z be a Poincaré space of formal dimension $n \geq 5$. let :

$$Z = Z_- \cup Z_+ \ , \quad Z_- \cap Z_+ = Z_0 \ , \quad \partial Z_\pm = \partial_\pm Z \cup Z_0, \quad \partial_\pm Z = Z_\pm \cap \partial Z$$

be a Q-decomposition of Z. Suppose that the homomorphisms

$$\pi_1(\partial Z_0) \to \pi_1(\partial_\pm Z) \ , \qquad \pi_1(Z_0) \to \pi_1(Z_\pm)$$

induced by the inclusions are isomorphisms. (If $\partial Z_0 = \emptyset$, only the second homomorphism is considered.) Then id_Z is Poincaré cobordant to $h : \bar{Z} \to Z$, so that the Poincaré space $\bar{Z}$ admits a Poincaré decomposition :

$$\bar{Z} = \bar{Z}_- \cup \bar{Z}_+ \ , \quad \bar{Z}_- \cap \bar{Z}_+ = \bar{Z}_0 \ , \quad \partial \bar{Z}_\pm = \partial_\pm \bar{Z} \cup \bar{Z}_0, \quad \partial_\pm \bar{Z} = \bar{Z}_\pm \cap \partial \bar{Z}$$

with $h(\bar{Z}_\pm) \subset Z_\pm$. Moreover, the space $\bar{Z}$ and its Poincaré decomposition may be chosen so that the homomorphisms :

$$\pi_1(\partial \bar{Z}_0) \to \pi_1(\partial_\pm \bar{Z}) \ , \qquad \pi_1(\bar{Z}_0) \to \pi_1(\bar{Z}_\pm)$$

induced by the inclusions are isomorphisms.

Proof : As $\pi_1(\partial Z_0) \to \pi_1(\partial_- Z)$ is an isomorphism, Theorem (4.4) implies that $id : \partial_- Z \to \partial_- Z$, which represents an element of $\Omega^Q_{n-1}(\partial_- Z, \partial Z_0)$ is in the image of $\Omega^P_{n-1}(\partial_- Z, \partial Z_0) \to \Omega^Q_{n-1}(\partial_- Z, \partial Z_0)$. Therefore, there is a Q-cobordism (W,V) from

$(\partial_- Z, \partial Z_0)$ to $(\partial_- \overline{Z}, \partial\overline{Z}_0)$, where $\partial_- \overline{Z}$ is a Poincaré space of formal dimension n-1 with $\partial(\partial_- \overline{Z}) = \partial\overline{Z}_0$, and a morphism of Q-spaces $G : (W,V) \longrightarrow (\partial_- Z, \partial Z_0)$ such that $G|\partial_- Z = \mathrm{id}$. (Observe that the spaces $\overline{Z}$ and $\overline{Z}_0$ do not yet exist; so far, we have only constructed $\partial_- \overline{Z}$ and $\partial\overline{Z}_0$.) Form the Q-space $W' = \partial Z \times [0,1] \cup W$. Its boundary is Q-decomposed : $\partial W' = V' \cup Y$, where :

$V' = V \cup \partial_+ Z$ is a Q-space and

$Y = (\partial Z \times 0) \cup \partial_- \overline{Z}$ is a Poincaré space.

Therefore, the map G represents an element of $\Omega^{QP}_{n+1}(\partial Z, \partial_+ Z)$. As $\pi_1(\partial_+ Z) \longrightarrow \pi_1(\partial Z)$ is an isomorphism (by the Van-Kampen Theorem), one has $\Omega^{QP}_{n+1}(\partial Z, \partial_+ Z) = 0$, by Theorem (4.4). Therefore, G is Q-cobordant (relative to Y) to $G : (\overline{W}, \overline{V}) \longrightarrow (\partial Z, \partial_+ Z)$, where $\overline{W}$ is a Poincaré space with $\partial\overline{W}$ Poincaré decomposed : $\partial\overline{W} = \overline{V} \cup Y$. Set $\partial\overline{Z} = \partial\overline{W} - Z\times 0$ and $\partial_+ \overline{Z} = \partial\overline{Z} - \mathrm{int}\,\partial_- \overline{Z}$. Then, we have already been able to replace the Q-decomposition of ∂Z, up to Poincaré cobordism, by a Poincaré decomposition.

As $\pi_1(Z_0) \longrightarrow \pi_1(Z_-)$ and $\pi_1(Z_-) \longrightarrow \pi_1(Z)$ are isomorphisms, the same argument as above relative to $\partial\overline{Z}$ permits us to replace in turn the Q-decomposition of Z, up to Poincaré cobordism, by a Poincaré decomposition $\overline{Z} = \overline{Z}_+ \cup \overline{Z}_-$, etc, which extends the one already obtained on $\partial\overline{Z}$. The conditions on the fundamental group are obtained by low dimensional Poincaré surgeries (using (3.16) and (3.18)). []

The second part of the above proof of (5.4) also shows the following statement :

(5.5) **Proposition** Let Z be a Poincaré space of formal dimension $n \geqslant 4$. let :

$$Z = Z_- \cup Z_+, \quad Z_- \cap Z_+ = Z_0, \quad \partial Z_\pm = \partial_\pm Z \cup Z_0, \quad \partial_\pm Z = Z_\pm \cap \partial Z$$

be a Q-decomposition of Z. Suppose that $\partial Z = \partial_- Z \cup \partial_+ Z$ is a Poincaré decomposition and that the homomorphisms $\pi_1(Z_0) \longrightarrow \pi_1(Z_\pm)$ induced by the inclusions are isomorphisms. Then id_Z is Poincaré cobordant, relative to ∂Z, to $h : \overline{Z} \longrightarrow Z$, so that the Poincaré space $\overline{Z}$ admits a Poincaré decomposition :

$$\overline{Z} = \overline{Z}_- \cup \overline{Z}_+ , \quad \overline{Z}_- \cap \overline{Z}_+ = \overline{Z}_0 , \quad \partial \overline{Z}_\pm = \partial_\pm \overline{Z} \cup \overline{Z}_0, \quad \partial_\pm \overline{Z} = \overline{Z}_\pm \cap \partial \overline{Z}$$

with $h(\overline{Z}_\pm) \subset Z_\pm$ and $h|\partial\overline{Z} = id$. Moreover, if $n \geqslant 5$, the space $\overline{Z}$ and its Poincaré decomposition may be chosen so that the homomorphisms $\pi_1(Z_0) \longrightarrow \pi_1(Z_\pm)$ induced by the inclusions are isomorphisms. []

<u>Proof of (5.3)</u> : Let $g : R \longrightarrow P$ represent an element of $\eta(P, \partial_+ P)$. By (2.15), There is a Poincaré decomposition of P :

$$P = P_1 \cup P_2, \quad P_1 \cap P_2 = P_0, \quad \partial P_i = \partial_i P \cup P_0,$$

$$\partial_- P \subset \operatorname{int} \partial_1 P, \quad \partial_1^+ P = \partial_1 P \cap \partial_+ P$$

such that :

a) P_2, $\partial_2 P$ and P_0 are manifolds

b) the pairs $(\partial_1^+ P, \partial P_0)$, (P_1, P_0), (P_2, P_0) and $(\partial_2 P, P_0)$ are 2-connected.

Replacing g by a Serre fibration and putting $R_i = g^{-1}(P_i)$ gives a decomposition of R :

$$R = R_1 \cup R_2, \quad R_1 \cap R_2 = R_0, \quad \partial R_i = \partial_i R \cup R_0, \quad \partial_- R \subset \mathrm{int}\partial_1 R,$$
$$\partial_1^+ R = \partial_1 R \cap \partial_+ R$$

and $g(R_i) = P_i$. This decomposition can be made into a Q-decomposition by (3.3). The pairs $(\partial_1^+ R, \partial R_0)$, (R_1, R_0), (R_2, R_0) and $(\partial_2 R, R_0)$ are 2-connected.

By Proposition (5.4), id_R is Poincaré bordant to $G : \overline{R} \to R$, where $\overline{R}$ admits a Poincaré decomposition :

$$\overline{R} = \overline{R}_1 \cup \overline{R}_2, \quad \overline{R}_1 \cap \overline{R}_2 = \overline{R}_0, \quad \partial \overline{R}_i = \partial_i \overline{R} \cup \overline{R}_0, \quad \partial_- \overline{R} \subset \mathrm{int}\partial_1 \overline{R},$$
$$\partial_1^+ \overline{R} = \partial_1 \overline{R} \cap \partial_+ \overline{R}$$

with $G(\overline{R}_i) \subset R_i$, and one may assume that $\pi_1(\partial R_0) \cong \pi_1(\partial_1^+ R)$, $\pi_1(\partial R_0) \cong \pi_1(\partial_2 R)$ and $\pi_1(R_0) \cong \pi_1(R_i)$ $(i = 1,2)$. Call $\overline{g} = g \circ G$ $: \overline{R} \to P$. It is then possible to perform a sequence of modifications on $\overline{g}$ and $\overline{R}$, up to Poincaré cobordism, according to the following steps :

<u>Step 1</u> : Using (5.2), (which is a consequence of Property 1 of (5.1)), change $\partial_1^+ \overline{g} = g | \partial_1^+ \overline{R} : \partial_1^+ \overline{R} \longrightarrow \partial_1^+ \overline{P}$, up to a

Poincaré cobordism relative to $\partial(\partial_- R)$, so that $\partial_1^+ g$ is a homotopy equivalence.

Step 2 : Again, using (5.2), change $\bar{g}_1 = g \circ R_1 : \bar{R}_1 \longrightarrow P_1$, up to a Poincaré cobordism relative to $\partial_1 \bar{R}$, so that $\bar{g}_1$ is a homotopy equivalence, as well as $\bar{g}_0 = \bar{g}_1 \vert \bar{R}_0 : \bar{R}_0 \longrightarrow P_0$.

Step 3 : as $j^{\bar{R}_2}$ admits a lifting through BO, use the classical $(\pi$-$\pi)$-theorem to change $\partial_2 \bar{R}$ into a manifold. (if $n = 6$, one uses the stable $(\pi$-$\pi)$-theorem, as in the proof of (4.17)).

Step 4 : Again, use the classical $(\pi$-$\pi)$-theorem to change $\bar{R}_2$ into a manifold, where $\partial \bar{R}_2 = \partial_2 \bar{R} \cup \bar{R}_0$ is a manifold decomposition, with $\partial_2 \bar{R}$ the manifold obtained in Step 3. Since this change is made by a homotopy equivalence, we still have that $\bar{g}_0$ is a homotopy equivalence.

By Property 2) of (5.1), we have $s(g) = s(\bar{g}) = t_1(s(\bar{g}_1)) + t_2(s(\bar{g}_2))$. By Property 1) of (5.1), we have $s(\bar{g}_1) = 0$, since $\bar{g}_1$ is a homotopy equivalence. The map $\bar{g}_2$ is a classical surgery problem. Therefore, by Property 3) of (5.1), $s(\bar{g}_2) = \sigma(\bar{g}_2)$, the Wall surgery obstruction. This proves that the value of $s(g)$ is completely determined by Properties 1), 2) and 3) of (5.1). Theorem (5.3) is thus proven. [ ]

B. MANIFOLD ENGULFINGS :

In this section, we introduce the material which will enable us to provide a geometric interpretation of the Poincaré surgery obstructions defined in Section 5.A. This material will be also used in Chapter 7.

Let P be a Poincaré space and let M be a connected smooth manifold. Let $f : (M,\partial M) \longrightarrow (P,\partial P)$ be a map (not supposed to be over BG). Suppose that there exists :

a) Poincaré decompositions : $\bar{P} = W \cup P_+$, $W \cap P_+ = P_0$, $\partial P = \partial_+ P \cup V$, where W is a compact smooth manifold with a manifold decomposition of its boundary : $\partial W = P_0 \cup V$, such that the pairs (P,W) and (P_+,P_0), $(\partial P,V)$ and $(\partial_+ P,\partial P_0)$ are 2-connected.

b) a homotopy equivalence $h : \bar{P} \longrightarrow P$.

c) a map $\bar{f} : (M,\partial M) \longrightarrow (W,V)$ such that $h \circ \bar{f}$ is homotopic to f.

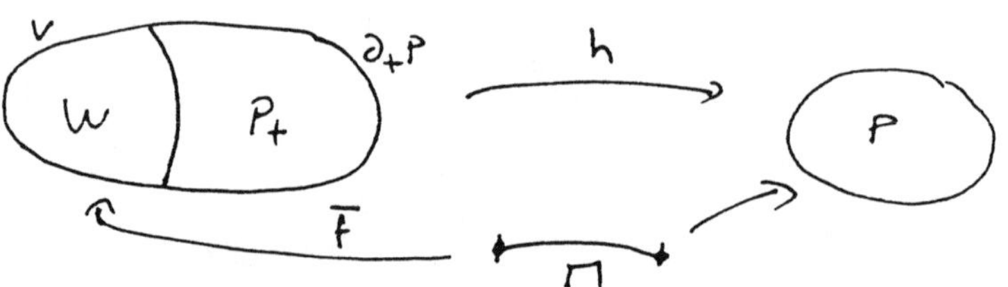

The 4-uple $(W,\bar{P},h,\bar{f})$ is called a **manifold engulfing of** f. In the case $\partial M = \emptyset$, one requires that $V = \emptyset$. Observe that $(V,\partial\bar{P},h|\partial\bar{P},\bar{f}|\partial M)$ is a manifold engulfing of $f|\partial M$. A 4-uple as above is called a **special manifold engulfing of** f if there is a homotopy of maps of pairs from f to a map having its image inside $(\partial W, \partial V)$.

Observe that a necessary condition for a map f to admit a manifold engulfing is that $j^P \circ f$ admits a lifting through BO. This condition is sometimes sufficient, as shown by the results of this section. We start with the case of a sphere.

(5.6) **Proposition** Let P be a Poincaré space of formal dimension $n \geq 5$. Let $f : S^j \longrightarrow P$ be a map with $j \leq [\frac{n+1}{2}]$ if $n \geq 6$, or $j \leq 2$ if $n = 5$, and such that $j^P \circ f : S^j \longrightarrow BG$ admits a lifting through BO Then, f admits a special manifold engulfing.

Proof of (5.6) : The proof divides into several cases :

Case 1 : n = 5 Take the decomposition given by (2.15) and make f cellular.

Case 2 : $n \geq 6$ By Proposition (2.15), there is a Poincaré decomposition :

$$P = T \cup P_1, \; T \cap P_1 = T'', \; T \cap \partial P = T',$$

where T is a manifold-thickening of a 2-dimensional complex, T' and T" are manifolds and the pairs (P_1, T''), (P,T) and $(\partial P \cap P_1, \partial T')$ are 2-connected.

Decompose S^j into two disks of dimension j : $S^j = D_- \cup D_+$. By general position, one can assume that

$f(D_-) = \{a\} \subset T''$ and $f(D_+) \subset P_1$.

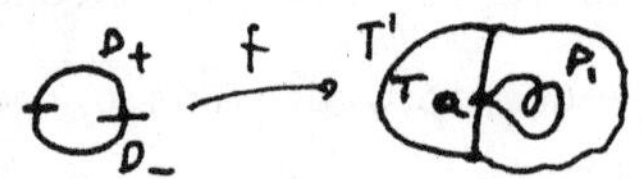

It is possible to assume that the image of D_- by the lifting of $j^P \circ f$ through BO coincides with $j^T(a)$, where $j^T : T \to BO$ is a lifting of $j^P | T$ (which exists since T is a manifold). Therefore, $j^T \vee (j^P \circ f) : T \vee S^j \to BG$ admits a lifting through BO.

The rest of the argument subdivides into two subcases :

<u>Subcase 2.a : $n \geq 7$</u> : Since P_1 is a Poincaré space, it is homotopy equivalent to a finite complex and therefore j^P : $P_1 \to BG$ admits a lifting through B_i for some i (see Section 4.C for the definition of B_i). By Lemma (4.24), Assertion $S(i,n)$ is true when $n \geq 7$. Therefore the map $f : (D_+, \partial D_+) \to (P_1, T'')$ can be realized by a Poincaré j-handle and $f|\partial D_+$ will extend to a map from D_- into the manifold T :

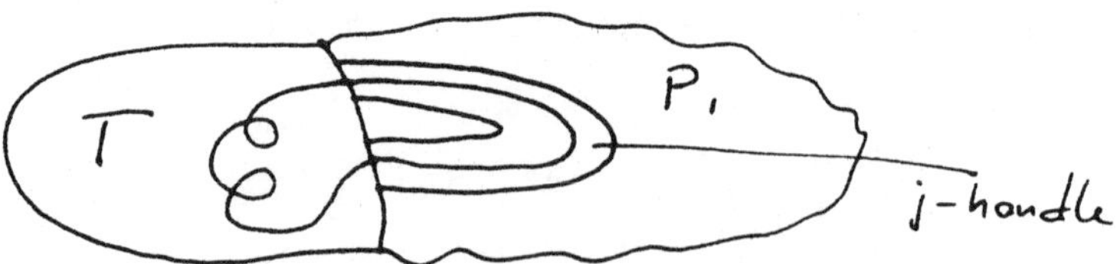

Set $W = T \cup$ (the above j-handle) and $P' = P - \text{int}W$. The spaces P' and W are Poincaré spaces, and thus $P = P' \cup W$ is a PD-decomposition which can be made into a Poincaré decomposition by (3.3). Observe that $\pi_1(\partial W) \cong \pi_1(W)$ and that

$j^W = j^P|W$ admits a lifting through BO. By the classical $(\pi\text{-}\pi)$-theorem, W has the homotopy type of a compact n-manifold W. The definition of f is obvious as well as the fact that the manifold engulfing obtained is a special manifold engulfing. Observe that this argument proves Proposition (5.6) in the case $n = 6$ and $\partial P = \emptyset$: instead of $S(i,n)$, we use $S_\emptyset(i,6)$, which is true by Lemma (4.24).

Subcase 2.b : n = 6 : Let K be the union of the handles of index ≤ 2 of a handle decomposition of T" starting with one 0-handle containing a. The boundary of P_1 is then Poincaré decomposed : $\partial P_1 = K \cup L$, and the pairs (P_1,L) and $(L,\partial L)$ are 2-connected. The proof of (4.17), which can be worked out relative to K, gives a homotopy equivalence $\alpha : \hat{P}_1 \longrightarrow P_1$ such that :

-. $\hat{P}_1$ has a Poincaré decomposition : $\hat{P}_1 = V_- \cup V_+$, $V_- \cap V_+ = V_0$, $V_\pm$ are manifolds with a manifold decomposition of their boundaries : $\partial V_\pm = \partial_\pm V \cup V_0$.

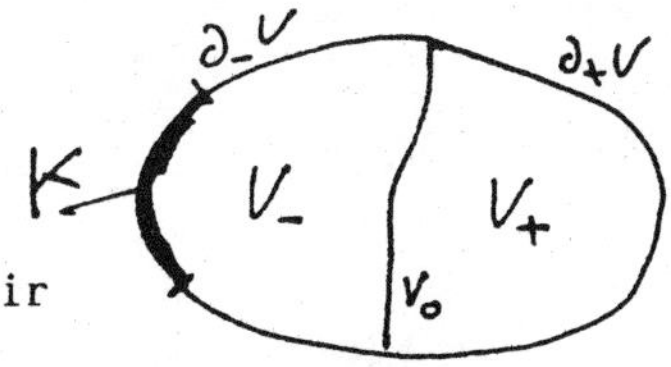

-. K is a codimension 0 submanifold of $\operatorname{int}\partial_- V$ and $\alpha|K = \mathrm{id}$.

The map $f|D_+$ is homotopic, relative to ∂D_+, to $\alpha\circ\bar{f}$, where $\bar{f}$ is a map from $(D_+,\partial D_+)$ to $(P_1,\{a\})$. The proof of (4.17) shows that $\bar{f}$ is homotopic (as a map from $(D_+,\partial D_+)$ to $(P_1,\partial_- V)$) to a map $\hat{f} : (D_+,\partial D_+) \longrightarrow (V_-,\partial_-V)$. By pushing this homotopy restricted to ∂D_+ into a collar neighborhood of ∂_-V in V_-, one can still assume that $\hat{f}(\partial D_-) = \{a\}$. Therefore, up to homotopy equivalence, $P = T \cup P_1 \cup \mathcal{M}_{\alpha|\partial P}$, and $\hat{f}$ can have image into V_- or $V_- \cup_K T$, which are codimension 0 submanifolds of P.

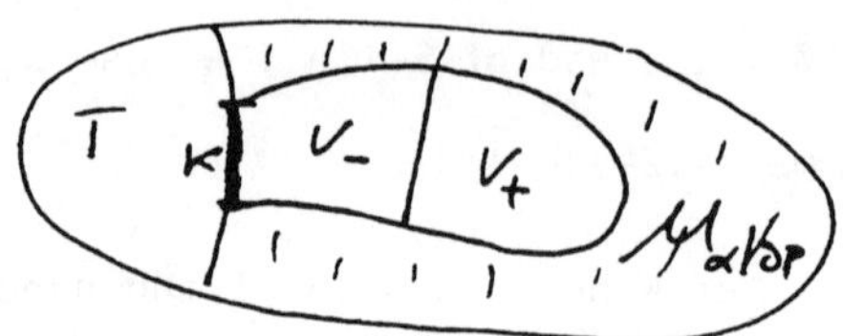

As the pair (V_-,K) is 2-connected, $\hat{f}|D_+$ can be realized by a 3-handle of (V_-,K) and one has the same situation as in subcase 2.a. []

A relative version of Proposition (5.8) is availble :

(5.7) **Proposition** Let P be a Poincaré space of formal dimension $n \geq 5$. Let $f : (D^j,S^{j-1}) \longrightarrow (P,\partial P)$ be a map, with $j \leq [\frac{n+1}{2}]$ if $n \geq 6$, or $j \leq 2$ if $n = 5$. Let $(V,L,h,\bar{f})$ be a manifold engulfing of $f|S^{j-1}$ (respectively, a special manifold engulfing of $f|S^{j-1}$) such that the lifting $j^V \circ \bar{f} : S^{j-1} \longrightarrow BO$ of $j^P \circ h \circ f$ extends to a lifting over D^j. Then, $(V,L,h,\bar{f})$ is induced by a manifold engulfing of f (respectively, a special manifold engulfing of f).

Proof : Case 1 : n = 5 As for (5.6), this follows from the results of Section 2.C.

<u>Case 2 : $n \geq 6$</u> Simplify the notation by replacing P by $P \cup \mu_h$, so one can assume that $L = \partial P$, $h = id$ and $\bar{f} = f$. By (2.20), there exists a Poincaré decomposition $P = T \cup P_1$, where T is a manifold obtained from V by adding 1 and 2-handles, $T \cap \partial P = V$, ∂T has a manifold decomposition $\partial T = V \cup T''$, where $T'' = (P_1 \cap T)$, and the pairs (P,T) and (P_1,T'') are 2-connected. By general position, there is a homotopy $F : S^{j-1} \times [0,1] \to T$ such that $F|S^{j-1} \times 0 = f$ and $F(S^{j-1} \times 1) \subset T''$. One may suppose that f coincides with F on a collar neighborhood C of S^{j-1} in D^j and that $f(D^j - C) \subset P_1$.

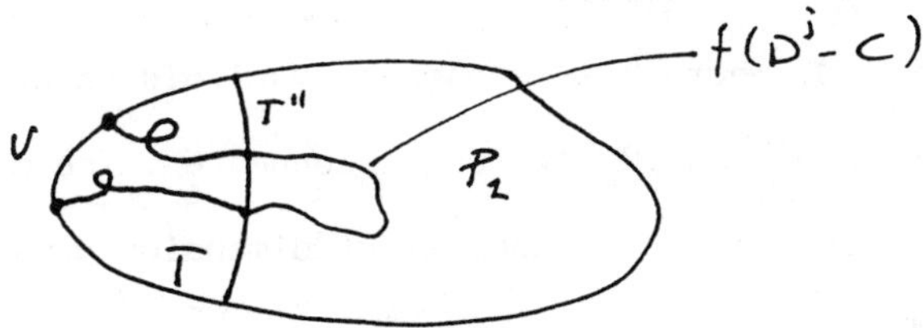

Let $f_1 = f|D^j - C$, considered as a map from (D^j, S^{j-1}) to (P_1, T''). By our hypotheses, the lifting of j^T through BO extends to a lifting of $j^P|T \cup_f D^j$. Therefore, as in the proof of (5.6), it is enough to realize f_1 by a Poincaré j-handle of (P_1,T'') and this realization is obtained exactly in the same way as in the proof of (5.1). []

We can now generalize Propositions (5.6) and (5.8) to manifolds :

(5.8) **Proposition** Let P be a Poincaré space of formal dimension $n \geq 5$. Let $f : M \longrightarrow P$ be a map, where M is a closed connected manifold of dimension $j \leq [\frac{n+1}{2}]$ if $n \geq 6$, or $j \leq 2$ if $n = 5$. Suppose that $j^P \circ f : M \longrightarrow BG$ admits a lifting through BO. Then, f admits a special manifold engulfing.

Proof : Let $M^{(0)} \subset M^{(1)} \subset \cdots \subset M^{(j)} = M$ be a handle decomposition of M. We shall prove the following assertion by induction on i:

Assertion E(i) : There exists :

a) a Poincaré decomposition $\overline{P} = W \cup P_+$, $W \cap P_+ = \partial W$, $W \cap \partial P = \emptyset$, where W is a compact manifold, such that the pairs (P,W) and $(P_+,\partial W)$ are 2-connected. The manifold W is a thickening of a complex of dimension $= \max\{2,i\}$ and $\pi_1(\partial W) \cong \pi_1(W)$.

b) a homotopy equivalence $h : \overline{P} \longrightarrow P$.

c) a map $\overline{f} : (M,M^{(i)}) \longrightarrow (P_+,\partial W)$ such that $h \circ \overline{f}$ is homotopic to f.

Assertion E(j) clearly implies Proposition (5.8). For Assertion E(0), one can take for W any smooth 2-skeleton of P given by (2.15). Let us then prove that E(i-1) implies E(i).

By induction on the number of i-handles of M, it is sufficient to prove that E(i-1) implies E(i) when there is only one handle of index i (if i = j, we assume that this is the case anyway). Write $f : (M,M^{(i-1)}) \longrightarrow (P_+,\partial W)$ for the map f obtained at the (i-1)-th stage. The restriction of f to the i-handle of M provides a map $f : (D^i,S^{j-1}) \longrightarrow (P_+,\partial W)$. Aplying (5.7) (case V = regular neighborhood of $M^{(i-1)}$ in ∂W), we can construct $\bar{P} = W \cup P_+$, etc, satisfying the above conditions a) and b) and a map $\hat{f} : M^{(i)} \longrightarrow \partial W$ such that $h \circ \hat{f}$ is homotopic to $f|M^{(i)}$. If i = j, we are finished. Otherwise, as h is a homotopy equivalence, $\hat{f}$ extends to $\bar{f} : M \longrightarrow P$ and, by general position, one may suppose that f(M) is contained in P_+. []

The same proof permits us to generalize (5.7) to the case where (D^i,S^{j-1}) is replaced by a manifold pair, obtaining the following statement (details are left to the reader) :

(5.9) **Proposition** Let P be a Poincaré space of formal dimension $n \geq 5$. Let $f : (M,\partial M) \longrightarrow (P,\partial P)$ be a map, where M is a compact manifold of dimension $j \leq [\frac{n+1}{2}]$ if $n \geq 6$, or $j \leq 2$ if n = 5. Let $(V,L,h,\bar{f})$ be a manifold engulfing of $f|\partial M$ (respectively, a special manifold engulfing of $f|\partial M$) such that the lifting $j^V \circ \bar{f} : \partial M \longrightarrow BO$ of $j^P \circ h \circ \bar{f}$ extends to a lifting over M. Then, $(V,L,h,\bar{f})$ is induced by a manifold engulfing of f (respectively, a special manifold engulfing of f).

We finally get the uniqueness result for manifold engulfings :

(5.10) **Proposition** Let P be a Poincaré space of formal dimension $n \geq 5$, or $n \geq 4$ if $\partial P = \emptyset$. Let $f_i : (M,\partial M) \longrightarrow (P,\partial P)$ be two homotopic maps ($i = 0,1$), where M is a compact manifold of dimension $j \leq [\frac{n}{2}]$ if $n \geq 5$, or $j \leq 1$ if $n = 4$. Let $\mathcal{E}_i = (W_i, \vec{P}_i, h_i, \vec{f}_i)$ be a manifold engulfings of f_i. (respectively, a special manifold engulfing of f_i). Let $F : (M\times[0,1], \partial M\times[0,1]) \longrightarrow (P\times[0,1], \partial P\times[0,1])$ be a homotopy between f_0 and f_1 so that the liftings of $j^P \circ f_i$ through BO determined by the manifold engulfing $\mathcal{E}_i$ extend to a lifting of $j^P \circ F$. Then, the manifold engulfings $\mathcal{E}_i$ are induced by a manifold engulfing of F (respectively, a special manifold engulfing of F).

Proof : This is just two successive applications of (5.9), the first one to induce the manifold engulfing of $f_0 \cup f_1|\partial M$ to a manifold engulfing of $F|\partial M\times[0,1]$ and the second to induce the resulting engulfing of $F|\partial(M\times[0,1])$ to a manifold engulfing of F. []

C. THE GEOMETRIC POINT OF VIEW

In this section, we show how to obtain the surgery results of Section 5.A in a geometric way, by actually performing surgeries on Poincaré spaces, following the presentation of [Wa2]. We start with the surgery below the middle dimension.

(5.11) <u>**Proposition**</u> **(Poincaré surgery below the middle-dimension)** Let P be a Poincaré space of formal dimension $n \geqslant 4$. Let $b \in \ker(\pi_j(j^P) : \pi_j(P) \longrightarrow \pi_j(BG))$, with $j < n/2$. Then a Poincaré surgery can be performed on b.

<u>Proof</u> : For n = 4 and 5, this result was already proven in Proposition (3.17). We then assume that $n \geqslant 6$.

Let $\beta : S^j \longrightarrow P$ reprsent b. As $j^P \circ \beta$ is homotopic to a constant map, it admits a lifting through BO. If $n \geqslant 5$, Proposition (5.6) makes it possible to have the image of β contained in a codimension 0 sub-manifold of P. By general position in manifolds, β is homotopic to a manifold-embedding which, using its normal bundle, determines a PD embedding. The result then follows from Proposition (3.15). []

The criterion for the possibility of performing a Poincaré surgery on $b \in \pi_k(P)$ for a Poincaré space of formal dimension 2k uses a self-intersection invariant μ, which generalizes those of [Wa2, Chapter5]. We present it here in the general set up which is needed in Section 7. Let $f : Y \longrightarrow X$ be a map over $\mathbf{R}P^\infty$ and let k be an integer. We define the abelian group $A(f,k)$ as the quotient of $Z\pi_1(X)$ by the sub-Z-module generated by the following sets :

- $\{x - f_*(a)x\overline{f_*(b)}\}$, $x \in Z\pi_1(X)$, $a,b \in \pi_1(Y)$,
- $\{x - (-1)^k\bar{x}\}$, $x \in Z\pi_1(X)$.

If X is a manifold of dimension 2k and Y a manifold of dimension k and f is an immersion in general position, the group $A(f,k)$ is the one into which the self-intersection number of f can be defined, as in [Wa2, Chapter5], and depends only on the regular class of the immersion.

We shall use the theory of immersions as in [Wa2], in the following adaptation to our framework :

(5.12) **Lemma** Let M^n and N^{n+k} be manifolds, considered as objects over BO using their stable normal bundle. Let $g : M \to N$ be a map over BO. Then, if $k \geq 1$, there exists an immersion $G : M\times D^k \longrightarrow N$, unique up to regular homotopy, such that $G|M\times 0$ is homotopic over BO to g.

Proof : The fact that g is over BO gives a stable vector bundle isomorphism :

$$\nu_M \xrightarrow{\cong} g^*\nu_N$$

Let τ_M and τ_N denote the stable tangent bundles. Let us extend by the identities the above isomorphism to an isomorphism :

$$\tau_M \oplus \nu_M \oplus g^*\tau_N \xrightarrow{\cong} \tau_M \oplus g^*\nu_N \oplus g^*\tau_N$$

Using the standard trivializations of $\tau_M \oplus \nu_M$ and of $g^*\tau_N \oplus g^*\nu_N$, one deduces an isomorphism between $\tau_M = \tau_{M\times D^k}$ and $g^*\tau_N$. Since $k > 1$, the existence and unicity of G follows from the classical immersion theory, as exposed in [Wa2, p. 10]. []

(5.13) **Proposition** Let P be a Poincaré space of formal dimension $n = 2k \geq 6$. Let $f : (M,\partial M) \longrightarrow (P,\partial P)$ be a pointed map over BG, where M is a k-dimensional compact manifold. Suppose that $f|\partial M$ is the restriction to ∂M of a Poincaré embedding of $\partial M\times D^k$. Suppose that $j^P\circ f$ admits a lifting j through BO, extending the lifting on ∂M given by the Poincaré embedding of $\partial M\times D^k$, and which classifies the stable normal bundle of M. Then there exists an element $\mu(f,j)\in A(f,k)$, depending only on the homotopy class of f and on j, such that :

1) $\mu(f,j) = 0$ if and only if f can be realized as the restriction to M of a Poincaré embedding of $M\times D^k$ extending those on $\partial M\times D^k$ and inducing the lifting j.

2) Suppose that f can be represented by an immersion β of $M\times D^k$ into a codimension 0 submanifold L of P, such that $j^L\circ\beta = j$. Then $\mu(f,j)$ is equal to the image of the self-intersection number $\mu(\beta) \in A(\beta,k)$ by the natural map $A(\beta,k) \longrightarrow A(f,k)$ induced by the inclusion $L \subset P$.

Proof : By adding 1 and 2-handle to $\partial M\times D^k$, one may suppose that $\partial M\times D^k$ is contained in a manifold engulfing of $f|\partial M$. Using the lifting j and Proposition (5.9), the manifold engulfing of $f|\partial M$ may be extended to a manifold engulfing $(W,\overline{P},h,\overline{f})$ of f. By the construction of these manifold engulfings in Section 5.B, it is clear that the map $\overline{f} : M \longrightarrow W$ is over BO. Since j classifies the stable normal bundle of M, Lemma (5.12) produces an immersion $G : M\times D^k \longrightarrow W$, whose regular homotopy class is unique. The self-intersection number $\mu(G)$ is then defined in $A(G,k) = A(f,k)$. Define $\mu(f,j)$ as the image of $\mu(G)$ in $A(f,k)$ under the natural map from $A(f,k)$ to $A(f,k)$ induced by the inclusion $W \subset \overline{P}$ and h. As the manifold engulfing $(W,\overline{P},h,\overline{f})$ is essentially unique by Proposition (5.10), the element $\mu(f,j)$ is well defined. If $\mu(f,j) = 0$, the standard Whitney trick permits us to eliminate the double points of the immersion. This proves 1).

To prove 2) add 1 and 2-handles to L so that we are dealing with a manifold engulfing. The conclusion of 2) is then clear from the above definition of $\mu(f,j)$. []

We are now in position to reproduce the geometric presentation of the surgery theory, following [Wa2]. Let $f : (P,\partial P) \longrightarrow (R,\partial R)$ be a morphism of Poincaré spaces of formal dimension $n = 2k \geq 6$, such that $f|\partial P : \partial P \longrightarrow \partial R$ is a homotopy equivalence.

As f is over BG, $\pi_j(f)$ is a subgroup of $\pi_j(j^P)$. Therefore, surgery below the middle dimension is possible for f (Proposition (5.11)) and f may be made k-connected, which means that its only non zero kernel is $K_k(f)$. As in [Wa2, Ch. 5], Poincaré duality implies that $K_k(f)$ is stably free and produces the intersection form λ. For the self-intersection form μ, let $b \in K_k(f) \subset \pi_{k+1}(j^P)$. Therefore b produces a commutative diagram :

$$\begin{array}{ccc} S^k & \subset & D^{k+1} \\ \downarrow \beta & & \downarrow \\ P & \xrightarrow{j^P} & BG \end{array} \qquad (5.14)$$

The map $D^{k+1} \longrightarrow BG$ may be seen as a hamotopy from a constant map to $j^P \circ \beta$. Make the map $BO \to BG$ a Serre fibration and use the lifting of the homotopies. This produces a lifting $j : S^k \longrightarrow BO$ of $j^P \circ \beta$. Using Proposition (5.13), we define $\mu(b) = \mu(\beta, j)$. Surgery can be performed on the classs b if and only if $\mu(b) = 0$. Hence, the triple $(K_k(f), \lambda, \mu)$ represents an element of $L_n(\pi_1(R); j^R)$ which plays exactly the same role as in [Wa2, Ch. 5].

When $n = 2k+1$, the map f can be made k-connected by surgery below the middle-dimension and the remaining non-zero kernels will be $K_k(f)$ and $K_{k+1}(f)$. Using Diagram (5.14), Proposition (5.6) and its proof and Lemma (5.12), a set of generators of

$K_k(f)$ may be represented by a disjoint union of embeddings of $S^k \times D^{k+1}$ into a codimension zero submanifold W of P, with $\pi_1(W) = \pi_1(P)$. By Proposition (5.10), two such realizations will be bordant by an immersion of $(\amalg S^k \times D^{k+1}) \times [0,1]$ into a codimension zero submmanifold of $P \times [0,1]$. Therefore the surgery obstruction $\sigma(f) \in L_n(\pi_1(R); j^R)$ can be defined and used as in [Wa2, Ch. 6].

6. HANDLE DECOMPOSITIONS FOR POINCARÉ SPACES

Let P be a Poincaré space of formal dimension n with a Poincaré decomposition $\partial P = \partial_- P \cup \partial_+ P$ of its boundary. A **handle decomposition** of $(P,\partial_- P)$ consists of :

a) a sequence of Poincaré spaces :

$$\partial_- P\times[0,1] = P_{(-1)} \subset P_{(0)} \subset P_{(1)} \subset \cdots \subset P_{(n)}$$

such that $P_{(j)}$ is obtained from $P_{(j-1)}$ by adding Poincaré handles of index j (see Section 3.D for the definition).

b) a homotopy equivalence $h : (P_{(n)},\partial P_{(n)},\partial_- P) \longrightarrow (P,\partial P,\partial_- P)$ extending the identity of $\partial_- P$.

When $\partial_- P = \emptyset$, we speak of a handle decomposition of P. Observe that, as in the manifold case, a handle decomposition of $(P,\partial_- P)$ determines a dual handle decomposition of $(P,\partial_+ P)$. The j-handles of the former correspond to the (n-j)-handles of the latter.

(6.1) **Theorem** Let P be a Poincaré space of formal dimension $n \geq 5$, with a Poincaré decomposition $\partial P = \partial_- P \cup \partial_+ P$ of its boundary. Then $(P,\partial_- P)$ admits a handle decomposition.

The proof of Theorem (6.1) makes use of the following lemma :

(6.2) **Lemma** Let P be a Poincaré space of formal dimension $n \geqslant 5$, with a Poincaré decomposition $\partial P = \partial_- P \cup \partial_+ P$ of its boundary. Suppose that $n \neq 6$ or $\partial_+ P = \emptyset$. Then, there exists a Poincaré decomposition :

$$P = P_- \cup P_+,\ P_- \cap P_+ = P_0,\ P_\pm \cap \partial P = \partial_\pm P$$

such that :

a) $(P_-, \partial_- P)$ admits a handle decomposition with handles of index $\leqslant [\frac{n}{2}]$.

b) The pair (P_+, P_0) is $[\frac{n}{2}]$-connected.

Proof : Case 1 : n = 5 : We take for P_- the union of $\partial_- P \times [0,1]$ with the manifold W of Proposition (2.15). Properties a) and b) are direct consequences of (2.15).

Case 2 : $n \geqslant 6$: For $k \leqslant [\frac{n}{2}]$, we shall prove the following assertion by induction :

Assertion **H**(k) : For $-1 \leqslant j \leqslant k$, there are Poincaré decompositions :

$$P = P_{(j)} \cup R_{(j)},\quad P_{(j)} \cap R_{(j)} = L_{(j)},\quad P_{(-1)} = \partial_- P \times [0,1]$$

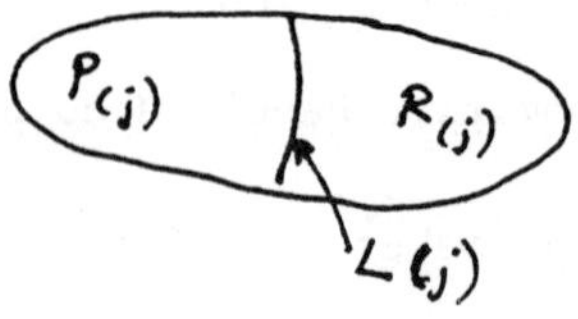

such that the pair $(R_{(j)},L_{(j)})$ is j-connected and $P_{(j)}$ is obtained from $P_{(j-1)}$ by adding Poincaré handles of index j.

Assertion **H**(2) can be easily obtained using Proposition (2.15). Suppose that **H**(k-1) is proven. As $(R_{(k-1)},L_{(k-1)})$ is a finite CW-pair, its first non-zero homotopy group $\pi_k(R_{(k-1)},L_{(k-1)})$ is a finitely generated $\mathbf{Z}\pi$-module ($\pi = \pi_1(P)$). We claim that each element of a finite set of generators of $\pi_k(R_{(k-1)},L_{(k-1)})$ can be realized by a Poincaré k-handle. Since $k \leq [\frac{n}{2}]$, this follows, when $n \geq 7$, from Assertion **S**(i,n) (see (4.19)), which is true for all i when $n \geq 7$ (See Proposition (4.24)). When n = 6, since then $\partial_+P = \emptyset$, one uses $\mathbf{S}_\emptyset(i,6)$ (proven in (4.20). The union of $P_{(k-1)}$ with these k-handles will be P(k), and their complement in $R_{(k-1)}$ will be $R_{(k)}$. The requirements for **H**(k) are easily verified. Therefore, **H**($[\frac{n}{2}]$) is proven by induction. This proves the lemma, setting $P_- = P_{([\frac{n}{2}])}$ and $P_+ = R_{([\frac{n}{2}])}$ []

<u>Proof of Theorem (6.1)</u> :

<u>Case 1 : n ≠ 6</u> : Take the Poincaré decomposition :

$$P = P_- \cup P_+,\ P_- \cap P_+ = P_0,\ P_\pm \cap \partial P = \partial_\pm P$$

given by Lemma (6.2). Take the Poincaré decomposition :

$$R = R_- \cup R_+,\ R_- \cap R_+ = R_0,\ R_\pm \cap \partial R = \partial_\pm R$$

given again by Lemma (6.2) for :

$$R = P_+,\ \ \partial_- R = \partial_+ P,\ \ \partial_+ R = P_0$$

Let $T = R_+$, which is a Poincaré cobordism between P_0 and R_0.

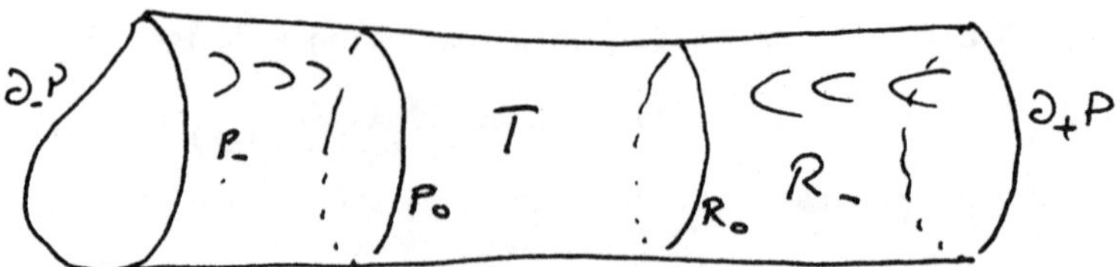

Since the pair (T,R_0) is $[\frac{n}{2}]$-connected, one has $H^i(T,R_0;Z\pi) =$ $= 0$ for $i \leq [\frac{n}{2}]$ ($\pi = \pi_1(T) = \pi_1(P)$). By Poincaré duality, one deduces that $H_i(T,P_0;Z\pi) = 0$ for $i \geq n - [\frac{n}{2}]$ and, since in addition (T,P_0) is $[\frac{n}{2}]$-connected, the inclusion $P_0 \subset T$ is a homotopy equivalence. The simmetric argument proves that the inclusion $R_0 \subset T$ is also a homotopy equivalence. Therefore T is a "Poincaré h-cobordism" and may be considered as the mapping cylinder of a homotopy equivalence $\alpha : R_0 \to P_0$. Using the handle decomposition of (R_-,R_0) dual to that of (R_-,∂_+P), this enables us to complete the handle decomposition of (P,∂_-P). (The handles attached to R_0 become, using α, Poincaré handles attached to P_0.)

<u>Case 2 : n = 6</u> : Let us first consider the case $\partial_+P = \emptyset$. As in Case 1, consider the Poincaré decomposition :

$$P = P_- \cup P_+,\ P_- \cap P_+ = P_0,\ P_\pm \cap \partial P = \partial_\pm P$$

given by Lemma (6.2). Change the notation : $R = P_+$, $\partial_-R = \emptyset$, $\partial_+R = P_0$. Take the Poincaré decomposition $R = W \cup T$,

$\partial T = \partial R \amalg \partial W$ which is given by (2.15) (case $V = \emptyset$), where W is a manifold with a handle decomposition with handles of index 2 and the pair $(T,\partial W)$ is 2-connected.

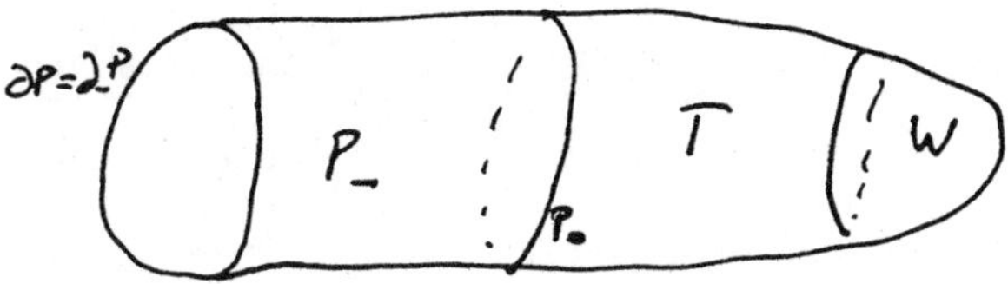

The Poincaré duality argument of Case 1 can be use here to prove that $(T,P_0,\partial W)$ is a Poincaré h-cobordism. The rest of the proof is then the same as in Case 1.

Theorem (6.1) is then true when $\partial_- P$ or $\partial_+ P$ is empty : if $\partial_+ P = \emptyset$, this is the above argument, and if $\partial_- P = \emptyset$, one uses the handle decomposition dual to that of $(P,\partial_+ P)$.

There remains the case where both $\partial_- P$ and $\partial_+ P$ are non-empty. Let $U = \partial_+ P$, and let :

$$\partial U \times [0,1] = U_{(-1)} \subset U_{(0)} \subset U_{(1)} \subset \cdots \subset U_{(5)} = U$$

be a Poincaré handle decomposition of U. By the above, the pair $(P,\partial P)$ admits a Poincaré handle decomposition :

$$\partial P \times [0,1] = P_{(-1)} \subset P_{(0)} \subset P_{(1)} \subset \cdots \subset P_{(6)} = P$$

The proof consists of working out a principle $\mathbf{GP}(j)$ of general position so that the j-handles of $P_{(j)}$ are attached on $\partial_- P$ $U_{(j)}$. It is then possible to construct a Poincaré handle decomposition $\{P_{(j)}\}$ of $(P,\partial_- P)$ by setting

$$P_{(j)} = P_{(j-1)} \cup \overline{\{U_{(j)}-U_{(j-1)}\}}\times[0,1] \cup \{j\text{-handles of } (P,\partial P)\}.$$

As $\partial_- P$ is not empty, **GP**(j) is easily obtained for $j \leq 2$, using the construction of $U_{(2)}$ and $P_{(2)}$ via the smoothings of 2-skeleta. This is also true for **GP**(3), since the attaching map of a 3-handle factors through a smooth 2-skeleton. By the construction of a handle decomposition for a Poincaré space of formal dimension 5, $V = U - \mathrm{int}U_{(2)}$ is a manifold-thickening of a 2 dimensional complex. Therefore, **GP**(4) follows from general position in V. As **GP**(5) and **GP**(6) are empty conditions, the proof is complete. []

(6.3) **Remarks**

a) Our handle decompositions are particular cases of L. Jones's "patch space structure" (see [Jo1]).

b) Nothing is known to us about existence of a Poincaré handle decomposition for a Poincaré space of formal dimension 4, even for a non-smooth closed topological 4-manifold.

c) Poincaré spaces of formal dimension 3 do not in general admit a handle decomposition. Indeed, some of them (with empty boundary) are not homotopy equivalent to closed manifolds (see [Wa4, p. 235]), and there is the following result :

(6.4) **Proposition** Let P be a Poincaré space of formal dimension 3 with a Poincaré decomposition $\partial P = \partial_- P \cup \partial_+ P$ of its boundary. Then, the pair $(P,\partial_- P)$ admits a handle decomposition if and only if P is homotopy equivalent to a 3-manifold.

<u>Proof</u> : let

$$\partial_- P\times[0,1] = P_{(-1)} \subset P_{(0)} \subset P_{(1)} \subset P_{(2)} \subset P_{(3)} = P$$

be a handle decomposition and let

$$\partial_+ P\times[0,1] = \overline{P}_{(-1)} \subset \overline{P}_{(0)} \subset \overline{P}_{(1)} \subset \overline{P}_{(2)} \subset \overline{P}_{(3)} = P$$

be its dual handle decomposition. By [EL], the spaces $\partial_\pm P$ are homotopy equivalent to surfaces and so one can assume that $P_{(-1)}$ and $\overline{P}_{(-1)}$ are manifolds. Poincaré 1-handles being smooth 1-handles, this implies that $\overline{P}_{(1)}$ and $\overline{P}_{(1)}$ are manifolds. Hence P is obtained by gluing $P_{(1)}$ to $\overline{P}_{(1)}$ via a homotopy equivalence $\alpha : \partial P_{(1)}-\partial_- P \longrightarrow \partial\overline{P}_{(1)}-\partial_+ P$. As homotopy equivalences of compact surfaces are homotopic to diffeomorphisms, it follows that P is homotopy equivalent to a 3-manifold. []

7. THE OBSTRUCTION TO TRANSVERSALITY

A. TRANSVERSALITY UP TO POINCARÉ BORDISM

Let (X,A) be a pair of spaces, with A closed in X. A **CD_q-structure** over (X,A) consists of a pair $(N_A, \partial N_A)$ of closed subspaces of X, giving rise to a decomposition :

$$X = N_A \cup [(X-N_A) \cup \partial N_A], \qquad N_A \cap [(X-N_A) \cup \partial N_A] = \partial N_A$$

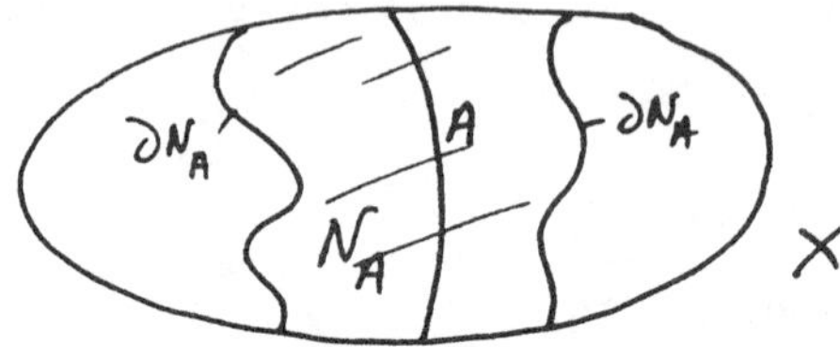

such that :

1) N_A is a neighborhood of A and the inclusion $A \subset N_A$ is a homotopy equivalence.

2) The theoretical fiber of the inclusion $\partial N_A \rightarrow N_A$ is homotopy equivalent to $S^{(q-1)}$. (We say that $\partial N_A \rightarrow N_A$ is a $(q-1)$-spherical fibration.)

3) The inclusion $\partial N_A \subset (N_A - A)$ is a homotopy equivalence.

The space N_A is called a **regular neighbourhood** of A (in X) and ∂N_A is called the **boundary** of N_A. To simplify the language in places, we might speak of a **CD_q-pair** (X,A) for a pair with a CD_q-structure, or of A being a **CD_q-subspace of** X. The letters CD_q stand for "codimension q".

Let (X,A) be a CD_q-pair and let P be a PD-space of formal dimension n. A map $f : P \to X$ is called **transverse to** A (with respect to the CD_q-structure of (X,A)) if the pair $(P,f^{-1}(A))$ admits a CD_q-structure such that :

a) $(N_{f^{-1}(A)}, \partial N_{f^{-1}(A)}) = (f^{-1}(N_A), f^{-1}(\partial N_A))$ and the associated decomposition is a PD-decomposition.

b) $f\ (N_{f^{-1}(A)}, \partial N_{f^{-1}(A)}) \to (N_A, \partial N_A)$ is a morphism of $(q-1)$-spherical fibrations (i.e. f induces a homotopy equivalence from the theoretical homotopy fiber of $\partial N_{f^{-1}(A)} \to N_{f^{-1}(A)}$ onto that of $\partial N_A \to N_A$.)

It follows from 1) that $f^{-1}(A)$ is a PD-space of formal dimension $n-q$ and that the inclusion $f^{-1}(A) \subset P$ is a PD-embedding.

In this section, an obstruction is defined for a map $f : P \to X$ as above to be PD-bordant (and, in the next section, to be homotopy equivalent) to a map transverse to A (see, for example Theorems (7.11), (7.12) and (7.23).) As in previous chapters, it will be convenient and more general to work in the category over BG and with Poincaré spaces instead of PD-spaces. We need a delicate notion of morphisms between CD_q-pairs which is the following : let (X,A) and (X',A') be CD_q-pairs, and let X and X' be spaces over BG. A map $f : X \to X'$ over BG is called a $\mathbf{CD_q}$**-map** if :

i) $f^{-1}(A') = f^{-1}(N_{A'}) \cap A$, $f^{-1}(N_{A'}) \subset N_A$, $f^{-1}(\partial N_{A'}) \subset \partial N_A$

ii) $(X, f^{-1}(A'))$ admits a CD_q-structure with

$$(N_{f^{-1}(A')}, \partial N_{f^{-1}(A')}) = (f^{-1}(N_{A'}), f^{-1}(\partial N_{A'})).$$

iii) The two maps $(f^{-1}(N_{A'}), f^{-1}(\partial N_{A'}))$ to $(N_A, \partial N_A)$ and to $(N_{A'}, \partial N_{A'})$ are morphisms of spherical fibrations.

It follows from i) that $f^{-1}(A')$ is open and closed in A. Observe that it is not required that $f(A) \subset A'$, and so f is not a map of pairs from (X,A) to (X',A'). In contrast, f does induce a map of pairs $(X, X-A) \longrightarrow (X', X'-A')$ and $(X, (X-N_A) \cup \partial N_A) \longrightarrow (X', (X'-N_{A'}) \cup \partial N_{A'})$. When $f : X \rightarrow X'$ is an inclusion, we say that (X,A) is a **CD_q-subpair** of (X',A') (again, note that a CD_q-subpair is not a subpair in the usual sense).

Let (X,A) be a CD_q-pair, with X over BG, and let P be a Poincaré space of formal dimension n. A map $f : P \rightarrow X$ is called **Poincaré-transverse to** A if :

a) f and $f|\partial P$ are CD_q-maps.

b) The inclusion $(\partial P, f^{-1}(A) \cap \partial P) \rightarrow (P, f^{-1}(A))$ is a CD_q-map.

c) The decomposition of P and ∂P given by the CD_q-structures are Poincaré decompositions. (In particular, f is PD-transverse to A.)

If P, in the above definition is just a Q-space, we say that f is **Q-transverse** if it satisfies Condition a) and b). The analog of condition c) is automatic : the decomposition of P and ∂P given by the CD_q-structures can be considered in a unique way as Q-decompositions, using (3.3).

We have already encountered particular cases of the following Q-transversality result :

(7.1) **Lemma** (Q-transversality) Let (X,A) be a CD_q-pair, with X over BG, and let Q be a Q-space. Then, any map $f : Q \longrightarrow X$ (over BG) is homotopy equivalent to a Q-transverse map to A.

Proof : Replace, using the standard procedure, the maps f and $f|\partial Q$ by Serre fibrations $\hat{f} : \hat{Q} \longrightarrow X$ and $\hat{f}|\widehat{\partial Q} : \widehat{\partial Q} \longrightarrow X$. (We use that the standard procedure guaranties that $\widehat{\partial Q} \subset \hat{Q}$ and then $\hat{Q}$ may be considered as a Q-space with $\partial\hat{Q} = \widehat{\partial Q}$.) One checks easily that f is Q-transverse to A. []

We now define new bordism groups by taking under consideration, in the definitions of $\Omega_n^P(X)$, $\Omega_n^Q(X)$ and $\Omega_n^{QP}(X)$, only maps from Poincaré spaces and Poincaré cobordisms (or Q-spaces and Q-coboridsms) which are Poincaré transverse (or Q-transverse) to A. We denote these groups by $\Omega_n^P(X;\pitchfork A)$, $\Omega_n^Q(X;\pitchfork A)$ and $\Omega_n^{QP}(X;\pitchfork A)$.

More generally, let (Y,B) be a CD_q-subpair of (X,A), that is (X,A) and (Y,B) are CD_q-pairs, Y is a a subspace of X over BG and the inclusion $Y \subset X$ is a CD_q-map (but again, B is not required to be a subspace of A). Let us consider, in the definitions of $\Omega_n^P(X,Y)$, only maps $f : (P,\partial P) \longrightarrow (X,Y)$ such that f is Poincaré transverse to A and $f|\partial P$ is Poincaré transverse to B (with the same restrictions on cobordisms). We then get a group $\Omega_n^P(X,Y;\pitchfork A,B)$. Analoguous definitions give rise to $\Omega_n^Q(X,Y;\pitchfork A,B)$ and $\Omega_n^{QP}(X,Y;\pitchfork A,B)$.

Even more generally, let (Y,B), (Z,C) and (T,D) be CD_q-subpairs of (X,A), with $T \subset Y \cap Z$. The same philosophy produces bordism groups :

$$\Omega_n^?\begin{pmatrix} T \subset Y & & D - B \\ \cap \quad \cap & ;\pitchfork & | \quad\; | \\ Z \subset X & & C - A \end{pmatrix}, \qquad ? = P,\ Q,\ QP.$$

which are related by the usual long exact sequence (see the proof of (7.3) or (7.5)). As a consequence of Lemma (7.1), we have :

(7.2) **Lemma** the forgetfull homomorphisms :

$$\Omega_n^Q(X;\pitchfork A) \longrightarrow \Omega_n^Q(X)$$

$$\Omega_n^Q(X,Y;\pitchfork A,B) \longrightarrow \Omega_n^Q(X,Y)$$

$$\Omega_n^Q\begin{pmatrix} T \subset Y & & D - B \\ \cap \quad \cap & ;\pitchfork & | \quad\; | \\ Z \subset X & & C - A \end{pmatrix} \longrightarrow \Omega_n^Q\begin{pmatrix} T \subset Y \\ \cap \quad \cap \\ Z \subset X \end{pmatrix}$$

are isomorphisms.

We can now start the theory of the obstruction to transversality up to Poincaré bordism.

(7.3) **Proposition** Let X be a space over BG, (X,A) be a CD_q-pair and (Y,B) a CD_q-subpair of (X,A). Let P be a Poincaré space of formal dimension n with a Poincaré decomposition of its boundary :

$$\partial P = \partial_- P \cup \partial_+ P\ , \qquad \partial_- P \cap \partial_+ P = \partial_0 P$$

Let $f : (P,\partial_+ P) \longrightarrow (X,Y)$ such that $f|(\partial_- P,\partial_0 P)$ is Poincaré transverse to A and B. Then, there is an obstruction w(f) in the group

$$T_n(X,Y;A,B) \overset{\text{def}}{=} \Omega^{QP}_{n+1}\begin{pmatrix} Y \subset X & & B - A \\ \cap \quad \cap & ;\pitchfork & | \qquad | \\ Y \subset X & & \emptyset - \emptyset \end{pmatrix}$$

such that f is Poincaré bordant, relative to $\partial_- P$ to a map which is Poincaré transverse to A and B if and only if $w(f) = 0$.

Proof : The map f represents an element

$$[f] \in \Omega^{P}_{n}\begin{pmatrix} Y \subset X & & B - A \\ \cap \quad \cap & ;\pitchfork & | \qquad | \\ Y \subset X & & \emptyset - \emptyset \end{pmatrix}$$

which is the obstruction for f to be Poincaré bordant, relative to $\partial_- P$ to a map which is Poincaré transverse to A and B. There is the long exact sequence :

$$\Omega^{QP}_{n+1}\begin{pmatrix} Y \subset X & & B - A \\ \cap \quad \cap & ;\pitchfork & | \qquad | \\ Y \subset X & & \emptyset - \emptyset \end{pmatrix} \xrightarrow{\ \partial\ } \Omega^{P}_{n}\begin{pmatrix} Y \subset X & & B - A \\ \cap \quad \cap & ;\pitchfork & | \qquad | \\ Y \subset X & & \emptyset - \emptyset \end{pmatrix} \longrightarrow$$

$$\longrightarrow \Omega^{Q}_{n}\begin{pmatrix} Y \subset X & & B - A \\ \cap \quad \cap & ;\pitchfork & | \qquad | \\ Y \subset X & & \emptyset - \emptyset \end{pmatrix} \longrightarrow$$

On the other hand, one has, using Lemma (7.2) :

$$\Omega_*^Q\begin{pmatrix} Y\subset X & & B - A \\ \cap \quad \cap & ;\pitchfork & | \qquad | \\ Y\subset X & & \emptyset - \emptyset \end{pmatrix} \xrightarrow{\cong} \Omega_*^Q\begin{pmatrix} Y\subset X \\ \cap \quad \cap \\ Y\subset X \end{pmatrix} = 0$$

Therefore, ∂ is an isomorphism and we set $w(f) = \partial^{-1}([f])$. []

Let (X,A) and (Y,B) as in (7.3). One then has :

$$Y\subset X, \quad B\subset Y, \quad A\subset X$$

$$A\cap Y\subset B, \quad N_A\cap Y = N_B, \quad \partial N_A\cap Y = \partial N_B$$

Moreover, $A\cap Y$ is open and closed in B.

(7.4) <u>CONVENTION</u> : The space $A\cup B$ shall be considered as a space over BG by taking the "Whitney sum" of $j^X|A\cup B$ with the classifying map of the fibration $\partial N_A\cup\partial N_B\longrightarrow A\cup B$.

(7.5) **<u>Proposition</u>** There is a long exact sequence :

$$\to T_n(X,Y;A,B)\longrightarrow \Omega^{QP}_{n-q}(A,A\cap Y)\oplus\Omega^{QP}_{n-q-1}(B-(A\cap Y))\longrightarrow$$

$$\longrightarrow \Omega^{QP}_n\begin{pmatrix} Y-B\subset X-A \\ \cap \qquad \cap \\ Y\subset X \end{pmatrix}\longrightarrow$$

<u>Proof</u> : There is a long exact sequence :

$$\to\Omega^{QP}_{n+1}\begin{pmatrix} Y\subset X & & B - A \\ \cap \quad \cap & ;\pitchfork & | \qquad | \\ Y\subset X & & \emptyset - \emptyset \end{pmatrix}\longrightarrow\Omega^{QP}_n\begin{pmatrix} Y-B\subset X-A & & \emptyset - \emptyset \\ \cap \qquad \cap & ;\pitchfork & | \qquad | \\ Y\subset X & & B - A \end{pmatrix}\longrightarrow$$

$$\longrightarrow\Omega^{QP}_n\begin{pmatrix} Y-B\subset X-A & & \emptyset - \emptyset \\ \cap \qquad \cap & ;\pitchfork & | \qquad | \\ Y\subset X & & \emptyset - \emptyset \end{pmatrix}\longrightarrow$$

The left-hand term is $T_n(X,Y;A,B)$ and the right-hand one is $\Omega_n^{QP}\begin{pmatrix} Y-B \subset X-A \\ \cap \quad \cap \\ Y \subset X \end{pmatrix}$. We then just have to identify the middle-term with $\Omega_{n-q}^{QP}(A,A \cap Y) \oplus \Omega_{n-q-1}^{QP}(B-(A \cap Y))$.

Let a be an element of $\Omega_n^{QP}\left(\begin{matrix} Y-B \subset X-A \\ \cap \quad \cap \\ Y \subset X \end{matrix} ; \pitchfork \begin{matrix} \emptyset - \emptyset \\ | \quad | \\ B - A \end{matrix}\right)$. It is represented by (Q,f), where :

-. Q is Q-space of formal dimension n

-. ∂Q is Q-decomposed :

$\partial Q = \bar{\partial} Q \cup \partial_- Q \cup \partial_+ Q$,

$\partial_- Q \cap \partial_+ Q = \partial_0 Q$,

$\partial_\pm Q \cap \bar{\partial} Q = \bar{\partial}_\pm Q$

$\partial_0 Q \cap \bar{\partial} Q = \partial(\partial_0 Q) = \partial(\bar{\partial}_\pm Q) = \bar{\partial}_0 Q$

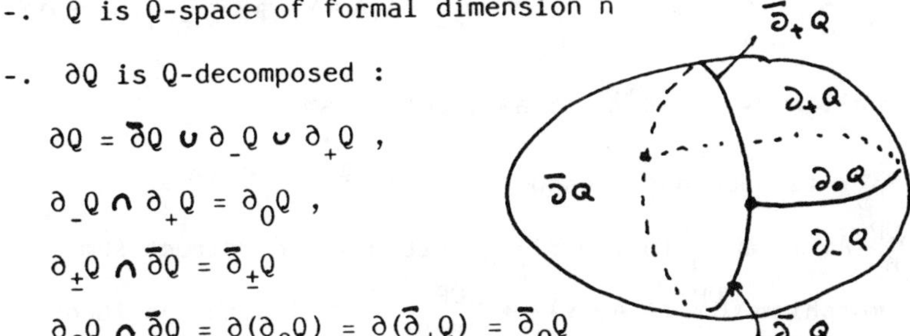

such that $\bar{\partial} Q$ is a Poincaré space of formal dimension n-1 and $\partial(\bar{\partial} Q) = \bar{\partial}_- Q \cup \bar{\partial}_+ Q$ is a Poincaré decomposition.

.- f is a map from $(Q,\partial_- Q,\partial_+ Q,\partial_0 Q)$ to $(X,X-A,Y,Y-B)$ whose restriction gives maps :

$f : (Q,\partial_+ Q) \longrightarrow (X,Y)$ Q-transverse to A and B and

$f : (\bar{\partial} Q,\bar{\partial}_+ Q) \longrightarrow (X,Y)$ Poincaré-transverse to A and B.

This definition implies that $f^{-1}(A)$ is a Q-space of formal dimension n-q, with Q-decomposed boundary $(f^{-1}(A) \cap \partial_+ Q) \cup (f^{-1}(A) \cap \bar{\partial} Q)$ and $(f^{-1}(A) \cap \bar{\partial} Q)$ is a Poincaré space of formal dimension n-q-1. We obtain this way an element of $\Omega_{n-q}^{QP}(A,A \cap Y)$

(the space A being over BG according to Convention (7.4)). Similarly, the space $f^{-1}(B-(A \cap Y)) \cap \partial_+ Q$ is a Q-space of formal dimension n-q-1, and, since $f(\partial_0 Q) \subset Y-B$, its boundary is a Poincaré space of formal dimension n-q-2. Using f, this gives an element of $\Omega^{QP}_{n-q-1}(B-(A \cap Y))$. These constructions extend to the cobordisms and give rise to a homomorphism :

$$\Omega^{QP}_n \begin{pmatrix} Y-B \subset X-A & & \emptyset - \emptyset \\ \cap \quad\quad \cap & ;\phi & | \quad\quad | \\ Y \subset X & & B - A \end{pmatrix} \xrightarrow{\mathcal{E}} \Omega^{QP}_{n-q}(A, A \cap Y) \oplus \Omega^{QP}_{n-q-1}(B-(A \cap Y))$$

We shall prove that $\mathcal{E}$ is an isomorphism.

As $A \subset Y$ is open and closed in B, one has $\Omega^{QP}_n(B) = \Omega^{QP}_n(A \cap Y) \oplus \Omega^{QP}_n(B-(A \cap Y))$. Therefore, the direct sum of the homomorphism $\Omega^{QP}_{n-q}(A, A \cap Y) \longrightarrow \Omega^{QP}_{n-q-1}(A \cap Y)$ with the identity of $\Omega^{QP}_{n-q-1}(B-(A \cap Y))$ fits in the long exact sequence :

$$\longrightarrow \Omega^{QP}_{n-q}(B) \longrightarrow \Omega^{QP}_{n-q}(A) \longrightarrow$$

$$\longrightarrow \Omega^{QP}_{n-q}(A, A \cap Y) \oplus \Omega^{QP}_{n-q-1}(B-(A \cap Y)) \longrightarrow \Omega^{QP}_{n-q-1}(B) \longrightarrow$$

One checks that the Diagram (7.5.a) below is commutative, where $\mathcal{E}_1$ and $\mathcal{E}_2$ are defined as $\mathcal{E}$. Let Q be a Q-space of formal dimension n-q, with ∂Q a Poincaré space, and let $g : Q \longrightarrow A$ represent an element in $\Omega^{QP}_{n-q}(A)$. Denote by ν the spherical fibration $\partial N_A \longrightarrow N_A$ and by $E(g^*\nu) \subset \hat{E}(g^*\nu)$ the total spaces of the induced sphere and disk-fibration over Q. Then $\hat{E}(g^*\nu)$ is a Q-space of formal dimension n, provided that

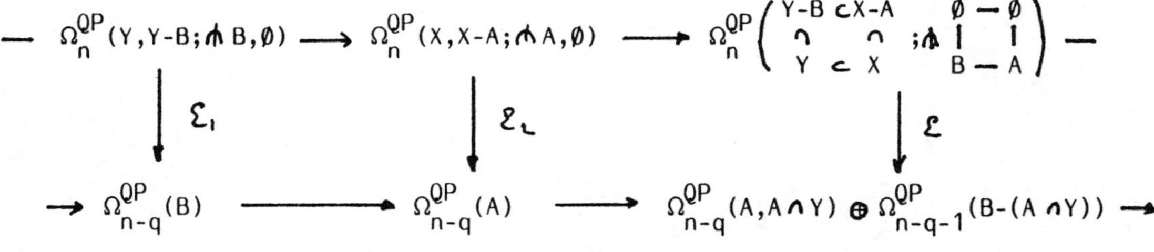
$\Omega_n^{QP}(Y, Y-B; \pitchfork B, \emptyset)$
$\Omega_n^{QP}(X, X-A; \pitchfork A, \emptyset)$
$\Omega_n^{QP}\begin{pmatrix} Y-B \subset X-A & ; \pitchfork & \emptyset - \emptyset \\ \cap \quad \cap & & | \quad | \\ Y \subset X & & B - A \end{pmatrix}$
$\mathcal{E}_1$
$\mathcal{E}_2$
$\mathcal{E}$
$\Omega_{n-q}^{QP}(B)$
$\Omega_{n-q}^{QP}(A)$
$\Omega_{n-q}^{QP}(A, A \cap Y) \oplus \Omega_{n-q-1}^{QP}(B-(A \cap Y))$

Diagram (7.5.a)

$\hat{E}(g^*\nu)$ is considered as a space over BG via the map

$$\hat{E}(g^*\nu) \longrightarrow Q \xrightarrow{g} A \longrightarrow X \xrightarrow{j^X} BG.$$

The boundary of $\hat{E}(g^*\nu)$ is Q-decomposed into $E(g^*\nu) \cup \hat{E}(g^*\nu|\partial Q)$ and $\hat{E}(g^*\nu|\partial Q)$ is a Poincaré space. The map g induces a map from $(\hat{E}(g^*\nu), E(g^*\nu))$ to $(X, X-A)$ which is Q-transverse to A. One checks that this construction gives rise to a homomorphism from $\Omega^{QP}_{n-q}(A)$ to $\Omega^{QP}_{n}(X, X-A; \pitchfork A, \emptyset)$ which is an inverse to $\mathcal{E}_1$. The same can be done for $\mathcal{E}_2$. Therefore, $\mathcal{E}_1$ and $\mathcal{E}_2$ are isomorphisms and, by the five lemma, $\mathcal{E}$ is an isomorphism. This proves Proposition (7.5). []

Observe that, if $q \geq 3$, then $\pi_1(X-A) \longrightarrow \pi_1(X)$ and $\pi_1(Y-B) \longrightarrow \pi_1(Y)$ are isomorphisms using Condition 3) of the definition of a CD_q-structure. Therefore, $\Omega^{QP}_n(X, X-A) \cong \Omega^{QP}_n(Y, Y-B) = 0$ for all n by Theorem (4.4) and then, for all n, one has

$$\Omega^{QP}_n \begin{pmatrix} Y-B \subset X-A \\ \cap \qquad \cap \\ Y \subset X \end{pmatrix} = 0$$

Hence, Proposition (7.5) has the following corollary :

(7.6) <u>**Corollary**</u> If $q \geq 3$, then

$$T_n(X, Y; A, B) \cong \Omega^{QP}_{n-q}(A, A \cap Y) \oplus \Omega^{QP}_{n-q-1}(B-(A \cap Y)) \qquad []$$

Theorem (4.4) permits us almost always to inject the groups $\Omega^{QP}_n(-)$ into surgery groups. Therefore, Proposition (7.3) together with Corollary (7.6) gives the following corollary :

(7.7) **Corollary** Let X be a space over BG, (X,A), (Y,B) and $f : (P,\partial_+ P) \longrightarrow (X,Y)$ be as in Proposition (7.3). Suppose that $q \geqslant 3$ and that either $n-q \neq 4$ or $A \cap Y = \emptyset$. Then there is an obstruction

$$w(f) \in L_{n-q-1}(\pi(A),\pi(A \cap Y);i^A) \oplus L_{n-q-2}(\pi(B-(A \cap Y));i^B)$$

such that f is Poincaré bordant to a map which is transverse to A and B if and only if $w(f) = 0$. Here $i^A : \pi(A) \longrightarrow \pi_1(BG)$ and $i^B : \pi(B) \longrightarrow \pi_1(BG) = \{\pm 1\}$ are induced by the maps from $A \cup B$ to BG considered in Convention (7.4). []

We now investigate what happens in codimension < 3.

(7.8) **Proposition** Let X be a space over BG, (X,A), (Y,B) and $f : (P,\partial_+ P) \longrightarrow (X,Y)$ be as in Proposition (7.3). Suppose that $q = 2$ and that the fibration $\partial N_A \cup \partial N_B \longrightarrow A \cup B$ is induced, for every connected component C of $A \cup B$, from a map $C \longrightarrow BG_2$ which is surjective on π_2. Then, $T_n(X,Y;A,B)$ is isomorphic to $\Omega^{QP}_{n-2}(A,A \cap Y) \oplus \Omega^{QP}_{n-3}(B-(A \cap Y))$.

Proof : The condition on the fibration $\partial N_A \cup \partial N_B \longrightarrow A \cup B$ implies that, for every connected component C of $A \cup B$, the map $\partial N_C \longrightarrow C$ induces an isomorphism on the fundamental group. Then, by Van-Kampen's theorem, both maps $X-A \subset X$ and $Y-B \subset Y$ induce an isomorphism on π_0 and π_1 for any base point. The argument is then the same as for Corollaries (7.6) and (7.7). []

For codimension 1, we take, for simplicity, the case $Y = \emptyset$.

(7.9) **Proposition** Suppose that $\partial N_A \subset N_A$ is an orientable S^0-fibration (i.e. A is bicollared in X). Then, there is an isomorphism :

$$T_n(X;A) \cong \Omega^{QP}_{n+1}\begin{pmatrix} \partial N_A & \subset & X-A \\ \cap & & \cap \\ N_A & \subset & X \end{pmatrix}$$

Moreover, if A cuts X into two pieces X_1 and X_2, then there is an isomorphism :

$$T_n(X;A) \cong \Omega^{QP}_{n+1}\begin{pmatrix} A & \subset & X_2 \\ \cap & & \cap \\ X_1 & \subset & X \end{pmatrix}$$

Proof : If $\partial N_A \to N_A$ is orientable, one has $(N_A,\partial N_A) = (A\times I, A\times\partial I)$, where $I = [0,1]$. The exact sequence of (7.5) becomes :

$$\to T_n(X;A) \to \Omega^{QP}_{n-1}(A) \xrightarrow{\alpha} \Omega^{QP}_n(A\times I, A\times\partial I) \longrightarrow$$

On the other hand, one has the exact sequence :

$$0 \longrightarrow \Omega^{QP}_n(A\times I, A\times\partial I) \xrightarrow{\partial} \Omega^{QP}_{n-1}(A) \oplus \Omega^{QP}_{n-1}(A) \xrightarrow{\beta} \Omega^{QP}_{n-1}(A) \longrightarrow 0$$

where β is the sum map. The homomorphism $\partial\circ\alpha$ is clearly the skew-diagonal map and therefore α is an isomorphism. This implies that $T_n(N_A,A) = 0$ and then :

$$T_n(X;A) \cong T_n(X,N_A;A,A)$$

Hence, the exact sequence of (7.5) produces the isomorphism :

$$\Omega^{QP}_{n+1}\begin{pmatrix} \partial N_A & \subset & X-A \\ \cap & & \cap \\ N_A & \subset & X \end{pmatrix} \xrightarrow{\cong} T_n(X;A) \qquad (1).$$

If A cuts X into two pieces X_1 and X_2, Formula (1) becomes :

$$T_n(X,A) \cong \Omega_{n+1}^{QP}\begin{pmatrix} A \amalg A & \subset & X_1 \amalg X_2 \\ \cap & & \cap \\ A & \subset & X \end{pmatrix}$$

On the other hand, the fact that $\Omega_*^{QP}(A,A) = 0$ implies the isomorphism $\Omega_*^{QP}(X_1 \amalg A, A \amalg A) \xrightarrow{\simeq} \Omega_*^{QP}(X_1, A)$ and then :

$$\Omega_*^{QP}\begin{pmatrix} A \amalg A & \subset & X_1 \amalg A \\ \cap & & \cap \\ A & \subset & X_1 \end{pmatrix} = 0$$

Hence :

$$T_n(X,A) \cong \Omega_{n+1}^{QP}\begin{pmatrix} A \amalg A & \subset & X_1 \amalg X_2 \\ \cap & & \cap \\ A & \subset & X \end{pmatrix} \xrightarrow{\simeq}$$

$$\xrightarrow{\simeq} \Omega_{n+1}^{QP}\begin{pmatrix} X_1 \amalg A & \subset & X_1 \amalg X_2 \\ \cap & & \cap \\ X_1 & \subset & X \end{pmatrix} \xrightarrow{\simeq} \Omega_{n+1}^{QP}\begin{pmatrix} A & \subset & X_2 \\ \cap & & \cap \\ X_1 & \subset & X \end{pmatrix} . \qquad []$$

(7.10) **Corollary** Suppose that A is connected and cuts X into two pieces X_1 and X_2, such that $\pi_1(A) \cong \pi_1(X_1)$. Then $T_n(X;A) = 0$.

Proof : By Van-Kampen's theorem, one has also $\pi_1(X_2) = \pi_1(X)$. The result then follows from the last statement of Proposition (7.9). []

Results (7.3) to (7.9) permits us to give answers to the question raised at the begining of this section. For instance, let (X,A) be a CD_q-pair (X not over BG), let P be a PD-space of formal dimension n, and $f : P \longrightarrow X$ be a map. The normal bundle of A determines a map $i^A : A \longrightarrow BG$ inducing $i^A : \pi_1(A) \rightarrow \pi_1(BG) = \{\pm 1\}$. In order to use our results, we have to consider the pair $(X\times BG, A\times BG)$, with $X\times BG$ over BG by the second projection. If we are dealing with oriented PD-spaces, one can use the pair $(X\times BSG, A\times BSG)$. Corollary (7.7) gives then the following result :

(7.11) **Proposition** Let (X,A) be a CD_q-pair (X not over BG), let P be a PD-space of formal dimension n, and $f : P \longrightarrow X$ be a map. Suppose that $q \geqslant 3$. Then, there is an element $W(f)$ $L_{n-q}(\pi(A)\times\{\pm 1\}; i^A \cdot id)$ so that f is PD-bordant to a map which is PD-transverse to A if and only if $W(f) = 0$.

(7.12) **Proposition** Let (X,A) be a CD_q-pair (X not over BG), let P be an oriented PD-space of formal dimension n, and $f : P \longrightarrow X$ be a map. Suppose that $q \geqslant 3$. Then, there is an element $W_+(f) \in L_{n-q}(\pi(A); i^A)$ so that f is oriented PD-bordant to a map which is PD-transverse to A if and only if $W_+(f) = 0$. The image of $W_+(f)$ in $L_{n-q}(\pi(A)\times\{\pm 1\}; i^A \cdot id)$ is $W(f)$.

Also, Corollary (7.10) gives :

(7.13) **Proposition** Let (X,A) be a CD_1-pair, such that A is connected and cuts X into two pieces X_1 and X_2, with $\pi_1(A) = \pi_1(X_1)$. Let P be a PD-space (respectively, an oriented PD-space) of formal dimension n, and $f : P \longrightarrow X$ be a map.

Then, f is PD-bordant (respectively, oriented PD-bordant) to a map which is PD-transverse to A.

B. TRANSVERSALITY UP TO HOMOTOPY EQUIVALENCE

Consider again a CD_q-pair (X,A), with X over BG, and a CD_q-subpair (Y,B) of (X,A). Let P be a Poincaré space of formal dimension n with a Poincaré decomposition of its boundary : $\partial P = \partial_- P \cup \partial_+ P$, $\partial_- P \cap \partial_+ P = \partial_0 P$. Let $f : (P,\partial_+ P) \to (X,Y)$ such that $f|(\partial_- P,\partial_0 P)$ is Poincaré transverse to A and B. We shall develop an obstruction theory for f to be homotopy equivalent (relative to $\partial_+ P$) to a map which is Poincaré transverse to A and B.

Replace, using the standard procedure, the maps $f : P \longrightarrow X$ and $f|\partial_+ P : \partial_+ P \longrightarrow Y$ by Serre fibrations. This gives a strictly commutative diagram :

$$\begin{array}{ccc} (P,\partial_+ P) & \xrightarrow{\bar{f}} & (\bar{X},\bar{Y}) \\ & \searrow^{f} & \downarrow \hat{f} \\ & & (X,Y) \end{array}$$

where $\hat{f}$ is a Serre fibration and $\bar{f}$ is a homotopy equivalence. Call $\bar{A} = \hat{f}^{-1}(A)$ and $\bar{B} = \hat{f}^{-1}(B)$. Then $(\bar{X},\bar{A})$ is a CD_q-pair and

$(\bar{Y},\bar{B})$ is a CD_q-subpair of $(\bar{X},\bar{A})$. The element $w(\bar{f}) \in T_n(\bar{X},\bar{Y};\bar{A},\bar{B})$ of Proposition (7.3) is defined, which depends only on the homotopy equivalence class of f and which vanishes if f is Poincaré transverse to A and B. Set $t(f) = w(\bar{f})$.

(7.14) **Theorem** Suppose that one of the following conditions holds :

1) $q \geq 3$
2) $q \leq 2$ and $n \geq 6$
3) $q \leq 2$, $n = 5$ and $\partial_+ P = \emptyset$.

Then, f is homotopy equivalent, relative to $\partial_- P$ to a map which is Poincaré transverse to (A,B) if and only if $t(f) = 0$.

The rest of this section is devoted to the proof of Theorem (7.14). Cases 2) and 3) are very different from Case 1), which is done first. In this case, the proof is linked to the proof of the following result :

(7.15) **Lemma** Let $g : R \longrightarrow Z$ be a morphism of Q-spaces of formal dimension $n \geq 5$, with $g|\partial R : \partial R \longrightarrow \partial Z$ a homotopy equivalence. Suppose that the spaces R, ∂R, Z and ∂Z have the homotopy type of finite complexes. Let $(A,B) \subset (Z,\partial Z)$ such that (Z,A) is a CD_q-pair with $q \geq 3$ and $(\partial Z,B)$ is a CD_q-subpair of (Z,A). Suppose that g is Q-transverse to (A,B), with B, $g^{-1}(B)$ and $g^{-1}(A)$ homotopy equivalent to finite complexes. Suppose that g is 2-connected, that

$H_*(g;Z\pi_1(Z)) = 0$ when $* \geq n-1$ and that $H^n(g;E) = 0$ for any $Z\pi_1(Z)$-module E. Then there exists a finite Q-space $\tilde{Z}$ with $\partial\tilde{Z} = \partial Z$ and a strictly commutative diagram :

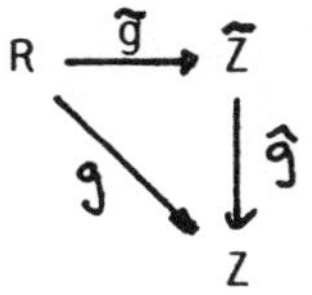

$$\begin{array}{ccc} R & \xrightarrow{\tilde{g}} & \tilde{Z} \\ & {\scriptstyle g}\searrow & \downarrow{\scriptstyle \hat{g}} \\ & & Z \end{array}$$

such that :

a) $\hat{g}$ is a homotopy equivalence

b) $\hat{g}|\partial Z = \mathrm{id}$, $\hat{g}$ is Q-transverse to (A,B). Let $\tilde{A} = \hat{g}^{-1}(A)$ and $\tilde{B} = \hat{g}^{-1}(B)$.

c) $\tilde{g}$ is Q-transverse to $(\tilde{A},\tilde{B})$, and the CD_q-structure on R for $\tilde{g}$ is the same as that for g.

d) $\hat{A}$ is homotopy equivalent to a finite complex.

(7.15.b) **<u>Remark</u>** If, in the statement of (7.15), one assumes that R is a Poincaré space and that g is Poincaré transverse to (A,B), it follows from c) that $\tilde{g}$ is Poincaré transverse to $(\tilde{A},\tilde{B})$.

To prove Theorem (7.14) in the case $q \geq 3$ and Lemma (7.15), we consider the following two assertions :

<u>Assertion T(m)</u> : Theorem (7.14) is true for $q \geq 3$ and $n \leq m$.

<u>Assertion F(m)</u> : Lemma (7.15) is true when $n \leq m$.

We shall prove below that :

a) T(5) is true.

b) T(m-1) implies F(m).

c) For $m \geq 5$, F(m) implies T(m).

This will prove Theorem (7.14) in the case $q \geq 3$ and Lemma (7.15) by induction on m.

Proof that T(5) is true : Observe that, by (7.6), $T_n(\bar{X},\bar{Y},\bar{A},\bar{B})$ is a sum of groups of the form $\Omega_*^{QP}(-)$ with $* \leq 2$. Therefore $T_n(\bar{X},\bar{Y},\bar{A},\bar{B}) = 0$ by (4.9.b). This explains why, in order to establish T(5), we shall prove the following lemma :

(7.16) **Lemma** Let f as in (7.14), with $n \leq 5$ and $q \geq 3$. Then f is homotopy equivalent relative to $\partial_- P$ to a map which is Poincaré transverse to (A,B).

Proof : Take a Poincaré decomposition $P = U \cup P_1$, where $U = \partial P \times [0,1]$. The decomposition $\partial P = \partial_- P \cup \partial_+ P$ extends to a Poincaré decomposition $U = U_- \cup U_+$, $U_- \cap U_+ = U_0$, with $U_- = \partial_- P \times [0,1]$ etc, which induces a Poincaré decomposition $\partial P_1 = \partial_- P_1 \cup \partial_+ P_1$, $\partial_- P_1 \cap \partial_+ P_1 = \partial_0 P_1$.

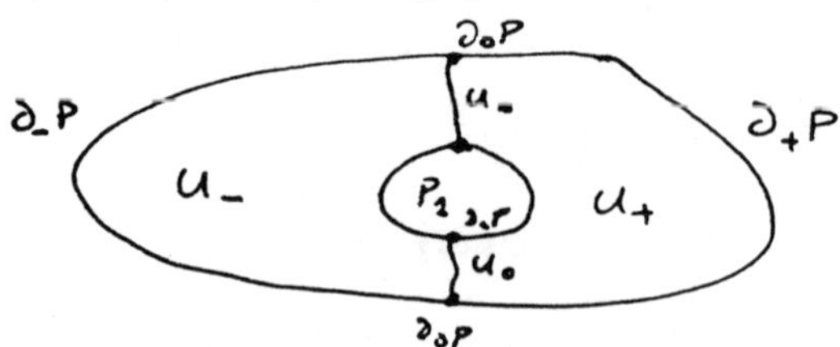

After a homotopy of f, one may suppose that

$$f(x,t) = f(x,0) \quad \text{for all } (x,t) \in \partial P \times [0,1]$$

Then $f|(U_-,U_0)$ is Poincaré transverse to (A,B).

Up to homotopy equivalence, one can always assume that ∂N_A and N_A are the total spaces of a $(q-1)$-sphere bundle ξ and its associated disk bundle. For a subspace R of A, one denotes by ∂N_R and N_R the corresponding total spaces for $\xi|R$. Observe that, if (A,R) is t-connected, then $(N_A, N_R \cup \partial N_A)$ is $(q+t)$-connected.

Let us define

$$K_0 = f^{-1}(B) \cap U_0 = K_0' \times [0,1]$$

$$K_- = f^{-1}(B) \cap U_- = K_-' \times [0,1]$$

The spaces K_0' and K_-' are Poincaré spaces of formal dimension $n-2-q \leq 0$ and $n-1-q \leq 1$. By cellularity, $f|K_-$ is homotopic, relative to K_-' to a map sending $K_0' \times \{1\}$ to the 0-skeleton $B^{(0)}$ of B and $K_-' \times \{1\}$ to the 1-skeleton $A^{(1)}$ of A. This homotopy can be extended to a homotopy of f to f' (relative to $\partial_- P$) such that $f'|U_-$ is Poincaré transverse to (A,B) and

$$f'(\partial_0 P_1) \subset \overline{Y - N_B} \cup N_{(B^{(0)})}$$

$$f'(\partial_- P) \subset \overline{X - N_A} \cup N_{(A^{(1)})}$$

$$f'(U_+) \subset Y$$

Define

$$X' = \overline{X - N_A} \cup N_{(A^{(2)})}$$

$$Y' = \overline{Y - N_B} \cup N_{(B^{(1)})}$$

The pairs $(X',A^{(2)})$ and $(Y',B^{(1)})$ are CD_q-subpairs of (X,A). The pair $(N_A, N_A(2) \cup \partial N_A)$ is $(q+2)$-connected and the pair $(N_B, N_B(1) \cup \partial N_B)$ is $(q+1)$-connected. Hence, there is no obstruction to producing a homotopy from f' to f'', relative to U_-, such that

$$f''(U_+) \subset Y, \qquad f''(P_1) \subset X', \qquad f''(\partial P_1) \subset Y'.$$

On the other hand, since $A^{(2)} \cup B^{(1)}$ is 2-dimensional, $\xi | A^{(2)} \cup B^{(1)}$ admits a linear reduction and thus, up to equivalence, one may suppose that $\xi | A^{(2)} \cup B^{(1)}$ is a vector bundle. By Propositions (2.15) and (2.16), there exists a Poincaré decomposition $P_1 = W \cup P'$, where W is a smooth manifold, such that

.- $\partial W = \partial_- W \cup \partial_+ W \cup (P' \cap W)$ (manifold decomposition)

.- $\partial_\pm W = \partial_\pm P_1 \cap W$, $\quad \partial_0 W = \partial_0 P_1 \cap W$

.- $\partial_0 W \supset f''^{-1}(N_B(1)) \cap \partial_0 P_1$, $\quad \partial_- W \supset f''(N_A(2)) \cap \partial_- P_1$

.- $\partial_+ W$ contains the $(n-q-1)$-skeleton of $\partial_+ P_1$

.- W " " $(n-q)$-skeleton of P_1

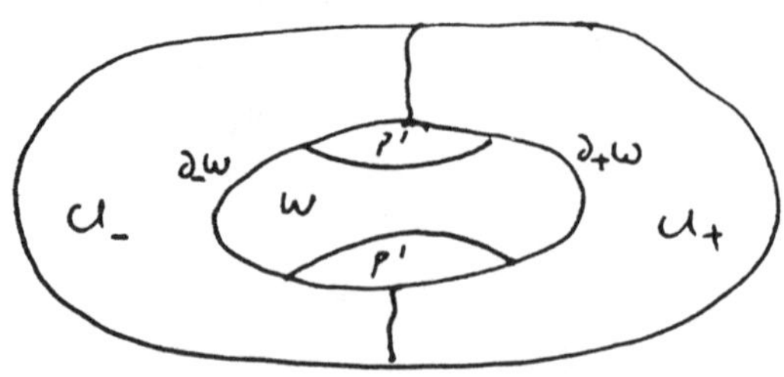

An easy computation involving Poincaré duality shows that

$$H^*(P' \cap \partial_- P_1, P' \cap \partial_0 P_1; \pi_*(Y', Y'-N_B)) = 0$$

$$H^*(P', P' \cap \partial_- P_1; \pi_*(X', X'-N_A)) = 0$$

Therefore, $f'' : (P_1, \partial_+ P_1) \longrightarrow (X', Y')$ is homotopic relative to $\partial_- P_1$ to a map f''' such that $(f'''^{-1}(A), f'''^{-1}(B)) = (f'''^{-1}(A^{(2)}), f'''^{-1}(B^{(1)})) \subset (W, \partial_+ W)$.

Since $f'''^{-1}(A) \cap \partial_- W = K'$ is a Poincaré space of formal dimension ≤ 1, one may as well assume that $f'''|\partial_- W$ is transverse regular in the smooth sense to $(A^{(2)}, B^{(1)})$. By transversality in the smooth category, a small perturbation of f''' (relative to $U_- \cup P'$) gives a map $g : (P, \partial_+ P) \longrightarrow (X, Y)$ such that $g|(W, \partial_+ W)$ is transverse regular to (A,B). This proves (7.16). []

<u>(7.17) Proof that T(m-1) implies F(m)</u> : Using the mapping cylinder of $g|\partial R$, one can identify ∂R with ∂Z and pretend that $g|\partial R = \mathrm{id}$. All the constructions are now done relative to ∂R and we can present them as if $\partial R = \emptyset$. Since $q \geq 3$, one may suppose that $n \geq 3$.

The conditions on R, Z and g imply that there exists a commutative diagram :

$$\begin{array}{ccc} R & \hookrightarrow & Z_1 \\ & {}_g\searrow & \downarrow\simeq \\ & & Z \end{array}$$

where Z_1 is obtained from R by adding finitely many cells of dimensions ≥ 2 and $\leq n-1$ (2-cells might be necessary only if $n = 3$, in which case there are attached trivially). The proof goes by induction on the number of these cells (one takes $Z = R$ and $g = id$ if this number is equal to 0). The first cell of (Z_1,R) is attached according to an element $a \in \pi_{i+1}(g)$. We shall establish the following affirmation :

<u>AFFIRMATION</u> : There exists a commutative diagram of morphisms of Q-spaces :

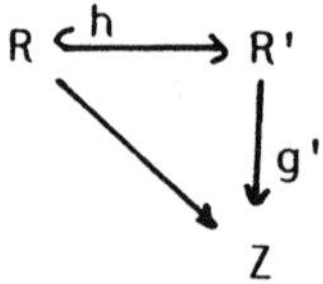

where :

-. g' is Q-transverse to A and $g'^{-1}(A) = A'$ is homotopy equivalent to a finite complex.

-. h is a homotopy equivalence which is Q-transverse to A' and the CD_q-structure on R for h is the same as that for g.

-. $h_*(a) \in \pi_{i+1}(g')$ is representable by a diagram :

$$\begin{array}{ccc} S^i & \xrightarrow{\alpha_S} & R' \\ \cap & & \downarrow g' \\ D^{i+1} & \xrightarrow{\alpha_D} & Z \end{array}$$

where α_S is Poincaré transverse to A' and α_D is Poincaré transverse to A.

This affirmation implies F(m). Indeed, set $R'' = R' \cup D^{i+1}$ and define $g'' : R'' \longrightarrow Z$ by extending g using α_D. The map g" is Q-transverse to A with

$$(g'')^{-1}(A) = g^{-1}(A) \cup \alpha_D^{-1}(A), \qquad \text{(union over } \alpha_S^{-1}(A))$$

Since α_D and α_S are Poincaré transverse to A and A', the space $(g'')^{-1}(A)$ is homotopy equivalent to a finite complex. The map g" enjoys the same properties as g but the number of cells of (Z,R") is strictly less that that of (Z,R). By induction hypothesis, $(\tilde{Z}'', \tilde{g}'', \hat{g}'')$ can be constructed for g" and we can take $\tilde{Z} = \tilde{Z}''$, $\tilde{g} = \tilde{g}''$ and $\hat{g} = \hat{g}''\, h$.

Proof of the affirmation : Write $C = g^{-1}(A)$. Replace Z by the mapping cylinder of g, so that one may consider that $R \subset Z$. Represent $a \in \pi_{i+1}(g)$ by a diagram :

$$a : \begin{pmatrix} S^i & \longrightarrow & R \\ \cap & & \cap g \\ D^{i+1} & \longrightarrow & Z \end{pmatrix}.$$

Since $i+1 \leq m-1$ and T(m-1) holds by hypothesis, the class a has a representative which is Poincaré transverse to (A,C) if and only if the obstruction $t(a) \in T_{i+1}(\bar{Z},\bar{R},\bar{A},\bar{C})$ vanishes. By (7.6), one has :

$$T_{i+1}(\bar{Z},\bar{R},\bar{A},\bar{C}) = \Omega^{QP}_{i+1-q}(\bar{A},\bar{C})$$

Therefore, $t(a) = 0$ if $i+1-q \leq 2$ by (4.9.b) and it remains to prove the case $i+1 \geqslant 5$.

Let $\hat{A}$ and $\hat{C}$ be the theoretical fiber of the maps $A \rightarrow Z$ and $C \rightarrow R$. One has

$$\begin{array}{ccccc} & \hat{A} & = & \hat{A} \\ & \downarrow & & \downarrow \\ & \bar{A} & \longrightarrow & A \\ & \downarrow & & \downarrow \\ D^{i+1} = & \bar{Z} & \longrightarrow & Z \end{array} \qquad\qquad \begin{array}{ccccc} & \hat{C} & = & \hat{C} \\ & \downarrow & & \downarrow \\ & \bar{C} & \longrightarrow & C \\ & \downarrow & & \downarrow \\ S^{i} = & \bar{R} & \longrightarrow & R \end{array}$$

Since $i+1 \geqslant 5$, the pairs $(\bar{A},\hat{A})$ and $(\bar{C},\hat{C})$ are 2-connected and therefore

$$\Omega^{QP}_{i+1-q}(\hat{A},\hat{C}) = \Omega^{QP}_{i+1-q}(\bar{A},\bar{C})$$

By Theorem (4.4), $\Omega^{QP}_{*}(\hat{A},\hat{C} \cup \hat{A}^{(2)}) = 0$ and Ω^{QP}_{*} commutes with inductive limits. Therefore, there exists a sequence :

$$\begin{array}{ccccccc} \hat{C} = E_0 & \subset & E_1 & \subset & E_2 & \subset & \cdots \\ & \searrow & \downarrow & \swarrow & & & \\ & & \hat{A} & & & & \end{array}$$

such that E_k is obtained by attaching to E_{k-1} a cell of dimension ≤ 2 and $t(a)$ has image zero in $\Omega^{QP}_{i+1-q}(\hat{A},E_k)$ for k big enough. We shall construct a sequence

$$R = R_0 \overset{h_1}{\hookrightarrow} R_1 \overset{h_2}{\hookrightarrow} R_2 \hookrightarrow \cdots \qquad g_0, g_1, g_2, \ldots : R_k \to Z \tag{7.18}$$

such that :

.- g_k is Q-transverse to A and $C_k = g_k^{-1}(A)$ is homotopy equivalent to a finite complex.

.- h_k is a homotopy equivalence Q-transverse to C_k.

.- There are inclusions

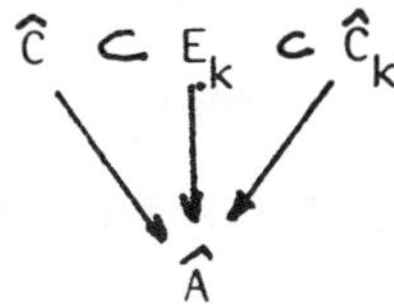

Therefore, for k big enough, one has $t(h_k \circ \ldots \circ h_1(a)) = 0$ in $\Omega^{QP}_{i+1-q}(\hat{A},\hat{C}_k) = T_{i+1}(\bar{Z},\bar{R}_k,\bar{A},\bar{C}_k)$. This will prove the affirmation, setting $h = h_k \circ \ldots \circ h_1$.

Sequence (7.18) is constructed by induction on k. Suppose that $R_1,\ldots,R_{k-1}$ have been constructed. The space E_k is obtained from E_{k-1} by attaching a cell according to an element $b^E \in \pi_s(\hat{A},E_{k-1})$, with $s \leq 2$, which has image b in $\pi_s(\hat{A},\hat{C}_{k-1})$. One has

$$\pi_s(\hat{A},\hat{C}_{k-1}) = \pi_{s+1}\begin{pmatrix} C_{k-1} & \longrightarrow & A \\ \downarrow & & \downarrow \\ R_{k-1} & \longrightarrow & Z \end{pmatrix}$$

and therefore, b can be represented by a diagram :

$$\beta \; : \; \begin{pmatrix} \beta_0 & \beta_- \\ & \\ \beta_+ & \beta \end{pmatrix} \; \vdots \; \begin{pmatrix} \partial_0 D & \subset & \partial_- D \\ \cap & & \cap \\ \partial_+ D & \subset & D \end{pmatrix} \longrightarrow \begin{pmatrix} C_{k-1} & \rightarrow & A \\ \downarrow & & \downarrow \\ R_{k-1} & \rightarrow & Z \end{pmatrix}$$

where $D = D^{s+1}$, $\partial_{\pm}D$ denote the upper and lower hemisphere of ∂D and $\partial_0 D$ its equator.

Let ν be the (q-1)-spherical normal bundle of A into Z (equivalent to the inclusion $\partial N_A \subset N_A$). Let N_- (respectively, N_0) be the total space of the disk bundle associated to $\beta_-^*(\nu)$ (respectively, to $\beta_0^*(\nu)$). Since $q \geqslant 3$, the inclusion $j : \partial_- D \rightarrow N_-$ can be deformed to an inclusion $j' : \partial_- D \subset \partial_{N_-}$. Also, β can be deformed to $\check{\beta}$ satisfying

.- $\mathrm{Im}\check{\beta}_+ \subset R_{k-1} - \mathrm{int}N_{C_{k-1}}$

.- $\mathrm{Im}\check{\beta}_0 \subset \partial N_{C_{k-1}}$

.- $\mathrm{Im}\check{\beta}_- \subset \partial N_A$

Also, since $s \leq 2$, Lemma (7.16) can be applied and one may suppose that $\check{\beta}$ is Poincaré transverse to A. Therefore, $\check{\beta}^{-1}(A)$ is a finite discrete set which is empty if $s+1 < 3$ or $q > 3$.

Form the space

$$W = N_- \cup_{j'} D$$

and define R_k as

$$R_k = R_{k-1} \cup W/\sim$$

where the equivalence relation $\sim$ is defined by

.- $x \in N_0 \subset W$ is identified with its image in N_{C_k}

.- $j(x) \sim \beta_-(x)$ for $x \in \partial_- D$.

In other words, one has attached to R_{k-1} the contractible space W over the contractible subspace $\partial_- D \cup N_0$. The inclusion $R_{k-1} \subset R_k$ is then a homotopy equivalence. The map g_k is defined by

$$g_k = \begin{cases} g_{k-1} & \text{on } R_{k-1} \\ \beta^{\gamma} & \text{on } D \end{cases}$$

It is immediate that

$$C_k = g_k^{-1}(A) = C_{k-1} \cup j(\partial_- D) \cup \beta^{\gamma-1}(A)$$

is a finite complex. The inclusion $C_{k-1} \subset C_k$ induces a map $\widehat{C}_{k-1} \to \widehat{C}_k$ and the element $b \in \pi_s(A, C_{k-1})$ goes to zero in $\pi_s(A, C_k)$ and, therefore, there is a factorization $C_{k-1} \subset E_k \subset C_k$. The other properties of (C_k, g_k) are easy to verify.

<u>(7.19) Proof that F(m) implies T(m) for m ⩾ 5</u> : Let f : $(P,\partial_+P) \longrightarrow (X,Y)$ as in Theorem (7.14). As T(5) is already established, one can assume that the formal dimension, n, of P is ⩾ 5. We first consider the case $\partial_+P = \emptyset$.

If $t(f) = w(\bar{f}) = 0$, it follows from (7.3) that there exists a Poincaré cobordism (W,P,R), trivial over ∂P, and a map $(F,\bar{f},g) : (W,P,R) \longrightarrow \bar{X}$ such that g is Poincaré transverse to $\bar{A}$.

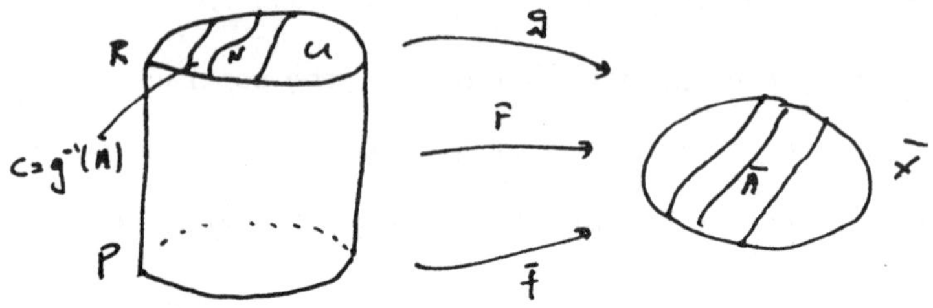

Let $C = g^{-1}(A)$ which is a Poincaré space of formal dimension n-q. Write $N = N_C$ and $U = R - \mathrm{int}C$. Observe that $\bar{X}$ is a Poincaré space of formal dimension n, since it is homotopy equivalent to P. By Poincaré surgeries in dimension ≤ 2, which may be done in U, one may assume that $\tilde{K}_i(R) = 0$ for i ≤ 2. Therefore, $H_*(g;Z\pi_1(X)) = 0$ for * ⩾ n-1 and $H^n(g;E) = 0$ for any $Z\pi_1(X)$-module E. Since T(m) is true by hypothesis, one can apply Lemma (7.15) together with Remark (7.15.b) and thus one may assume that $\bar{A}$ is a finite complex.

Observe that, by the (π-π)-theorem (5.2), it is possible to

perform Poincaré surgeries on W and R so that W becomes an h-cobordism and g and F homotopy equivalences. The idea is to perform the surgeries on R in such a way to keep the fact that g is Poincaré transverse to $\overline{A}$.

Perform first Poincaré surgeries on C (suitably thickened to N) so that

$$\widetilde{K}_i(C) = \widetilde{K}_i(N) = 0 \quad \text{for } i \leq \frac{n-q-2}{2}$$

By the excision and Thom isomorphisms, one deduces that

$$\widetilde{K}_i(R,U) = \widetilde{K}_i(N,\partial N) = 0 \quad \text{for } i \leq \frac{n+q-2}{2}$$

Observe that, as $\overline{X}$, $N_{\overline{A}}$ and $\partial N_{\overline{A}}$ are finite complexes, the space $\overline{X}$-int$N_{\overline{A}}$ is homotopy equivalent to a finite complex (all these spaces have same fundamental group, since $q \geq 3$). Therefore, one can perform Poincaré surgeries on U such that

$$\widetilde{K}_i(U) = 0 \quad \text{for } i \leq \frac{n-2}{2}$$

Using the long exact sequence

$$\to \widetilde{K}_i(U) \longrightarrow \widetilde{K}_i(R) \longrightarrow \widetilde{K}_i(R,U) \longrightarrow \widetilde{K}_{i-1}(U) \longrightarrow$$

one sees that $\widetilde{K}_i(R) = 0$ for $i \leq \frac{n-2}{2}$. It is also possible to perform Poincaré surgeries on intW so that

$$\widetilde{K}_i(W) = 0 \quad \text{for } i \leq \frac{n-1}{2}$$

Now, the classical proof of the $(\pi$-$\pi)$theorem, which can be achieved in Poincaré spaces using the technique of Section

5.C, shows that one can perform Poincaré surgeries in dimension $k = [n/2]$ on R and in dimension k+1 on W to transform F and g into homotopy equivalences $\check{F}$ and $\check{g}$. As $\tilde{K}_k(U) \cong \tilde{K}_k(R)$ and $\pi_1(U) \cong \pi_1(R)$, the surgeries on R may be performed inside intU. Therefore, the existence of $\check{F}$ makes $\hat{f} \circ \check{g}$ homotopy equivalent to f and $\hat{f} \circ \check{g}$ is Poincaré transverse to $\overline{A}$. We have established T(m) in the absolute case $(Y = \emptyset)$.

The general case is decomposed into two successive absolute cases in which we apply the above technique. Since T(5) is established one may suppose that $n \geq 6$. The fact that $t(f) = 0$ produces as above a map $(W,P,R) \longrightarrow \vec{X}$, but the Poincaré cobordism W induces a Poincaré cobordism $(V,\partial_+P,\partial_+R)$ and g is Poincaré transverse to $(\overline{A},\overline{B})$. We first proceed as above with $F|V$ (since $\partial_+(\pi_+R)$ is empty; observe that ∂_+P is of formal dimension ≥ 5), which enables us to assume that $g|\partial_+R$ is Poincaré transverse to $\overline{B}$. We then do the same with F relative to $V \cup \partial_-P\times[0,1]$.

<u>(7.20) Proof of Theorem (7.14), Cases 2) and 3)</u> : As in (7.19), the fact that $t(f) = w(\vec{f}) = 0$ implies that there exists Poincaré cobordisms $(W,P,R) \supset (V,\partial_+P,\partial_+R)$ and a map $(F,\vec{f},g) : (W,P,R) \longrightarrow \vec{X}$ such that g is Poincaré transverse to $(\overline{A},\overline{B})$. The idea is to perform Poincaré surgeries on these Poincaré spaces to make W and V h-cobordisms from (P,∂_+P),

preserving the transversality properties of g. As in (7.19), it is enough to prove (7.14) in the case $q \leq 2$, $n \geq 5$ and $\partial_+ P = \emptyset$; Case 2) reduces to two successive applications of this particular case.

As $q \leq 2$, one may assume that $(N_{\bar{A}}, \partial N_{\bar{A}})$ and $(N_C, \partial N_C)$ are pairs of total spaces of disk and sphere bundles associated to vector bundles over $\bar{A}$ and C. Therefore, it makes sense to say that a map from a manifold to $\bar{X}$ is smoothly transverse to $\bar{A}$ (see [Br1, Section I.2]). We shall use the following result :

(7.21) **Lemma** Let M be a manifold of dimension $i \leq n/2$. Let $\alpha : M \longrightarrow R$ be a map such that $j^R \circ \alpha$ admits a lifting through BO. Then, there is a manifold engulfing $(T, \bar{R}, h, \bar{\alpha})$ for α (see Section 5.B) such that $g \circ h$ is Poincaré transverse to $\bar{A}$ and $g \circ h | T$ is smoothly transverse to $\bar{A}$.

We defer the proof of (7.21) to the end of this section and show now how (7.21) enables us to finish the proof of Theorem (7.14).

By the $(\pi$-$\pi)$-theorem, one knows that (F,g) is Poincaré bordant to homotopy equivalences $(\check{F}, \check{g})$. Using Section 5.C, one knows that $\check{g}$ is obtained from g by a sequence of surgeries in the following way : one has a sequence of maps $g_i : R_i \longrightarrow \bar{X}$, with $g_0 = g$ and $g_{[n/2]} = g$. The map g_{i+1} is obtained from g_i by surgeries along a smooth embeddings

$\alpha_i : \amalg (S^i \times D^{n-i}) \longrightarrow T_i$, where T_i is a codimension 0 sub-manifold of R obtained by manifold engulfing of some map of $\amalg(S^i \times D^{n-i})$ to R_i.

Suppose that g_i is Poincaré transverse to $\bar{A}$. One can then use a manifold engulfing (T_i, α_i) of the type of Lemma (7.21) (we identify $\bar{R}_i$ with R_i, so that h_i = id, to save notations). Make the embedding $\alpha_i : \amalg S^i \times 0 \longrightarrow T_i$ smoothly transverse to $C_i = g_i^{-1}(\bar{A})$ (by the conclusion of (7.21), $C_i \subset T_i$ is a codimension 2 smooth submanifold of T_i). Therefore $g_i \circ \alpha_i$ is smoothly transverse to $\bar{A}$. As α_i is used to perform surgeries, $g_i \circ \alpha_i$ extends to $\beta_i : \amalg D^{i+1} \longrightarrow \bar{X}$, which can be chosen smoothly transverse to $\bar{A}$. Write $(K_i, \partial K_i) = (g_i \ \alpha_i^{-1}(\bar{A}), \alpha_i^{-1}(C_i))$, a codimension 2 sub-manifold pair of (D^{i+1}, S^i). One may suppose that the inverse image of C_i is $S^i \times D^{n-i}$ for the embedding α_i is $\partial K_i \times D^{n-i}$.

It is now clear that surgeries along this α_i produce a map g_{i+1} which is Poincaré transverse to A with

$$C_{i+1} = [C_i - \mathrm{int}\partial K_i \times D^{n-i}] \cup K_i \times S^{n-i-1}.$$

It remains now to prove Lemma (7.21).

Let $C = g^{-1}(A)$, which is a Poincaré space of formal dimension n-q. Make α smoothly transverse to C. Let $K = \beta^{-1}(C)$, which is a codimension 2 submanifold of M. By Proposition (5.8), $\alpha | K$

admits a manifold engulfing, i.e., one may suppose that $\alpha(K)$ is contained in a codimension 0 submanifold L of C (when dimC = 3 (or 4), we just use the existence of smoothings of the 0 (or 1) skeleton of C guaranteed by Section 2.C). Then N_L is a smooth codimension 0 submanifold of N_C which contains the image of an embedded regular neighborhood N_K of K in M. By Proposition (5.9), there exists a manifold engulfing T_0 for $\alpha|M - N_K : M - N_K \longrightarrow R - N_C$, such that $T_0 \cap \partial N_C = \partial N_L$. Set T = $T_0 \cup N_L$. Observe that $g|T$ is smoothly transverse to $\bar{A}$.

The manifold T constitutes a manifold engulfing of α, except in the cases $(n,q) \in \{(5,1),(5,2),(6,2)\}$, where the pair (R,T) is not 2-connected. To obtain a manifold engulfing in these cases, one adds handles of index 1 and 2 to T in R - T, which can be taken smoothly transverse to C, as done above for the manifold M. []

As for (7.11) to (7.13), the results of Section 7.B permits us to answer to the second question raised at the begining of Section 7.A : let (X,A) be a CD_q-pair (X not over BG), let P be a PD-space of formal dimension n, and $f : P \longrightarrow X$ be a map; when is f homotopy equivalent to a map which is PD-transverse to A ?

Recall that the normal bundle of A determines a map $i^A : A \longrightarrow BG$ inducing $i^A : \pi_1(A) \longrightarrow \pi_1(BG) = \{\pm 1\}$. In order to use our results, we have to consider the pair $(X\times BG, A\times BG)$, with $X\times BG$ over BG by the second projection. If we are dealing with oriented PD-spaces, one can use the pair $(X\times BSG, A\times BSG)$. The framework for using Theorem (7.14) is then the following one

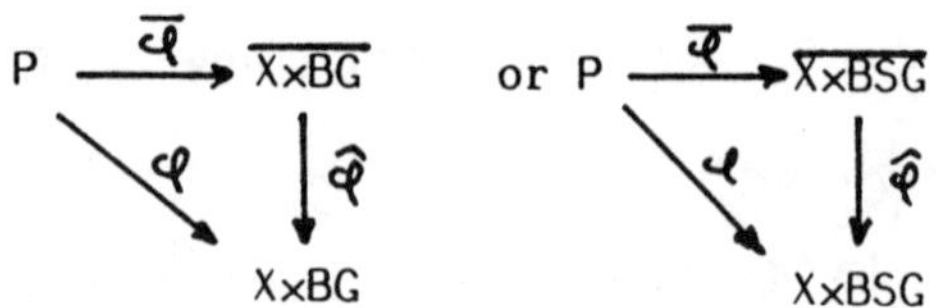

where $\varphi = (f,\mu)$, with μ classifying the Spivak bundle of P. By Theorem (7.14), the obstruction for f being homotopy equivalent to a map which is PD-transverse to A belongs to $T_n(\overline{X\times BG}, \overline{A\times BG})$, where $\overline{A\times BG} = \hat{\varphi}^{-1}(A\times BG)$, and the obstruction for f being oriented homotopy equivalent to a map which is PD-transverse to A belongs to $T_n(\overline{X\times BSG}, \overline{A\times BSG})$. We give below two particular cases of interest.

(7.22) **Proposition** Let (X,A) be a CD_1-pair, such that A is connected and cuts X into two pieces X_1 and X_2, such that the pairs (X_i,A) are 2-connected. Let P be a PD-space (respectively, an oriented PD-space) of formal dimension $n \geq 5$ (or $n \geq 6$ if $\partial P \neq \emptyset$), and $f : P \longrightarrow X$ be a map. Then, f is homotopy equivalent (respectively, oriented homotopy equivalent) to a map which is PD-transverse to A.

Proof : Since (X_i, A) are 2-connected, then $\pi_1(\overline{A\times BG}) = \pi_1(\overline{X_i\times BG})$, and, by Corollary (7.10), $T_n(\overline{X\times BG}, \overline{A\times BG}) \cong T_n(\overline{X\times BSG}, \overline{A\times BSG}) = 0$. Proposition (7.22) then follows from Theorem (7.14). []

When $q \geqslant 3$, Theorem (7.14) and Corollary (7.7) permit us to express our obstruction as an element of $L_{n-q}(\pi(A\times BG); (i^A \times id)\circ \mathfrak{a})$. If the map f is 2-connected, $\pi(A\times BG)$ may be computed and one has

(7.23) **Proposition** Let (X,A) be a CD_q-pair (X not over BG), let P be a PD-space of formal dimension n, and $f : P \longrightarrow X$ be a 2-connected map. Suppose that $q \geqslant 3$. Then, there is an element $T(f) \in L_{n-q}(\pi(A); i^A \times j)$, where j is the composed map

$$j : \pi(A) \longrightarrow \pi(X) \xrightarrow{f_*^{-1}} \pi(P) \longrightarrow \pi_1(BG) = \{\pm 1\}$$

and f is homotopy equivalent to a map which is PD-transverse to A if and only if $T(f) = 0$.

Proof : It is enough to establish that the pair $(\pi(\overline{A\times BG}), (i^A \times Id)\circ \hat{\mathfrak{a}})$ is isomorphic to the pair $(\pi(A), i_A \times j)$. By considering each component of A, one may suppose that A is connected.

Let F and F' be the fiber of the maps $X\times BG \xrightarrow{pr_1 \circ \mathfrak{a}} X$ and $X\times BG \xrightarrow{\mathfrak{a}} X\times BG$. One has a morphism of Serre fibrations :

$$\begin{array}{ccc}
F' & \longrightarrow & F' \\
\downarrow & & \downarrow \\
F & \longrightarrow & \overline{A\times BG} \\
\downarrow & & \downarrow \hat{\ell} \\
BG & \longrightarrow & A\times BG
\end{array}$$

Since f is 2-connected, F is 1-connected and, in the homotopy exact squence of therefore

- $\pi_2(BG) \longrightarrow \pi_1(F')$ is surjective
- $\pi_1(BG) \longrightarrow \pi_0(F')$ is an isomorphism.

Hence, $\pi_2(A\times BG) \longrightarrow \pi_1(F')$ is surjective and one has a short exact sequence

$$0 \longrightarrow \pi_1(\overline{A\times BG}) \longrightarrow \pi_1(A\times BG) \longrightarrow \pi_0(F')$$

The following diagram is cartesian and co-cartesian :

$$\begin{array}{ccc}
\pi_1(A\times BG) & \xrightarrow{pr_1 \circ \hat{\ell}_*} & \pi_1(BG) \\
\downarrow pr_1 & & \downarrow \\
\pi_1(A) & \xrightarrow{\partial} & \pi_0(F')
\end{array}$$

The pair $(\pi_1(\overline{A\times BG}), pr_1 \circ \hat{\ell}_*)$ is isomorphic to $(\pi_1(A), \partial)$ and the map ∂ from $\pi_1(A)$ to $\pi_0(F') = \pi_1(BG) = \{\pm 1\}$ is nothing else but j. Hence $(\pi(\overline{A\times BG}), (i^A \times Id)\circ \hat{\ell})$ is isomorphic $(\pi(A), i_A \times j)$. []

8. THE QUINN EXACT SEQUENCE FOR OTHER KIND OF POINCARE SPACES

A. SIMPLE POINCARÉ SPACES

A **simple Poincaré space** is a Q-space which is a Poincaré complex in the sense of [Wa2, Chapter 2]. All the results of Chapters 2 to 7 work under the same hypotheses for simple Poincaré spaces. One has just to proceed to make the following changes :

Poincaré space	Simple Poincaré space
Homotopy equivalence	Simple homotopy equivalence
$L_n = L_n^h$	L_n^s

The proofs are exactly the same, except that, each time that a finitely generated free $Z\pi$-module occur, it has to based. A special mention may be deserved for the result of Eckmann-Mueller, which is used in the proof of Proposition (4.8) : this result says that if P is a finite Poincaré space of formal dimension 2, then there is a homotopy equivalence f between P and a surface. But f is actually a simple homotopy equivalence since the Whitehead group of the fundamental group of a surface vanishes, by a classical theorem of Waldhausen [Wd].

B. Λ-POINCARE SPACES

Let (X,Y) be a pair over BG. Let Λ be a ring with anti-involution. We suppose that there is given a homomorphism of rings with anti-involution $Z\pi_1(X) \longrightarrow \Lambda$.

Define the nth **Λ-Poincaré bordism group** $\Omega_n^{\Lambda P}(X,Y)$ of (X,Y) as the group of bordism classes of pairs (P,f), where $f : (P,\partial P) \to (X,Y)$ is a map over BG and P is a Λ-Poincaré space of formal dimension n. The definition of a Λ-Poincaré space is the following one :

- P is a Q-space of formal dimension n which is a $(\Lambda\text{-PD})$-space (see Section 2.A) with $t((P)) = [P]$.
- P has a smooth 2-skeleton, in the sense of Proposition (2.15).

The definition of a Λ-Poincaré bordism is given accordingly, requiring that the smooth 2-skeleta are smoothly cobordant.

<u>**Remark**</u> : Suppose that for any homomorphism $G \longrightarrow \pi_1(X)$, the ring homomorphism $ZG \longrightarrow \Lambda$ is locally epic (for instance, if Λ is a ring between **Z** and **Q**). Then, for $n \geqslant 5$, the existence of a smooth 2-skeleton is guaranteed by Proposition (2.15) (and (2.20) for the corresponding condition on cobordisms).

As in Section 4, there is the exact sequence :

$$\longrightarrow \Omega_{n+1}^{Q\Lambda P}(X,Y) \longrightarrow \Omega_n^{\Lambda P}(X,Y) \longrightarrow \Omega_n^{Q}(X,Y) \longrightarrow \Omega_n^{Q\Lambda P}(X,Y) \longrightarrow$$

where the relative bordism groups $\Omega_n^{Q\Lambda P}(X,Y)$ are defined as for the groups $\Omega_n^P(X,Y)$, but using Λ-Poincaré spaces. As in Chapter 4, we would like to identify these groups with surgery groups, obtaining a Quinn exact sequence. The relevant surgery groups are the Cappell-Shaneson groups $\Gamma_n(\mathcal{F})$ (see [CS1]), where $\mathcal{F}$ is the ring homomorphism diagram :

$$\mathcal{F} = \begin{pmatrix} Z\pi_1(Y) & \longrightarrow & \Lambda \\ \downarrow & & \| \\ Z\pi_1(X) & \longrightarrow & \Lambda \end{pmatrix} \qquad \text{if } Y \neq \emptyset \text{ ;}$$

$$\mathcal{F} = (Z\pi_1(X) \longrightarrow \Lambda) \text{ ,} \qquad \text{if } Y = \emptyset.$$

For this we need the usual assumption that the homomorphism $Z\pi_1(X) \to \Lambda$ is locally epic (see Section 2.C). The precise result is the following :

(8.1) **Theorem** Suppose that $Z\pi_1(X) \longrightarrow \Lambda$ is locally epic. Then, for $n \geqslant 7$ (or $n = 6$ and $Y = \emptyset$), there is an isomorphism :

$$\sigma_n : \Omega_{n+1}^{Q\Lambda P}(X,Y) \longrightarrow \Gamma_n(\mathcal{F})$$

which is natural and commutes with boundary homomorphisms.

Remark : a) A more general definition could be taken by considering in addition, for each connected component Y_i of Y, a ring with anti-involution Λ_i and a commutative diagram of locally epic homomorphisms of rings with anti-involution :

$$\mathcal{F} : \begin{pmatrix} \coprod Z\pi_1(Y_i) & \longrightarrow & \coprod \Lambda_i \\ \downarrow & & \downarrow \\ Z\pi_1(X) & \longrightarrow & \Lambda \end{pmatrix}$$

The surgery groups for the corresponding bordism group $\Omega^{Q\mathfrak{F}P}_{n+1}$ is then the Cappell-Shaneson groups $\Gamma_n(\mathfrak{F})$.

b) The Cappell-Shaneson groups can be replaced by Wall groups of some localized group ring, using the results of [Vo]. The condition locally epic is also used.

The proof of (8.1) needs the following two lemmas, whose easy proof are left to the reader.

(8.2.a) **Lemma** $\Gamma_n(\mathfrak{F}) = \lim \Gamma_n(\mathfrak{F}_\alpha)$, where the inductive limit is taken over all the homomorphisms $\alpha : A \longrightarrow \pi_1(X)$ where A a finitely presented group, and $\mathfrak{F}_\alpha : ZA \longrightarrow \Lambda$ is the ring homomorphism obtained by composing α with $\mathfrak{F}$. []

(8.2.b) **Lemma** Suppose that $Z\pi_1(X) \longrightarrow \Lambda$ is locally epic. Let $\alpha : A \longrightarrow \pi_1(X)$ be a homomorphism with A finitely presented. Let $a_1,\ldots,a_k \in \Lambda$. Then there exist a factorization of α

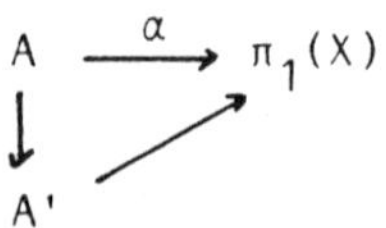

where A' is finitely presented, and a unit u of Λ so that $\{ua_1,\ldots,ua_k\}$ is in the image of $ZA' \longrightarrow \Lambda$. []

Proof of (8.1) : Suppose first that $Y = \emptyset$ and that j^X admits a lifting $\tilde{j}^X$ through BO. A homomorphism $\sigma_n = \sigma_n(j^X)$: $\Omega^{Q\Lambda P}_{n+1}(X,Y) \longrightarrow \Gamma_n(\mathfrak{F})$ is then defined for $n \geqslant 5$, as in Proposition (4.7) and will be a composed homomorphism :

$$\Omega^{Q\wedge P}_{n+1}(X) \longrightarrow \Omega^{\wedge P}_{n}(X) \xrightarrow{\overline{\sigma}_n} \Gamma_n(\mathcal{F}) .$$

To define $\overline{\sigma}_n = (\overline{\sigma}_n(\hat{j}^X))$, let (P,f) represent an element of $\Omega^{\wedge P}_n(X)$ and ξ be a stable vector bundle over P with characteristic map $\hat{j}^X \circ f : P \longrightarrow BO$. The pair $(\xi,(P))$ determines a surgery problem

$$\begin{array}{ccc} \nu_M & \longrightarrow & \xi \\ \downarrow & & \downarrow \\ (M,\partial M) & \xrightarrow{g} & (P,\partial P) \end{array}$$

with associated Cappell-Shaneson surgery obstruction $\sigma(g)$ $\Gamma_n(Z\pi_1(P) \to \Lambda)$. We define $\overline{\sigma}_n(P,f) = f_*(\sigma(g))$. As in the proof of Porposition (4.7), one checks that $\overline{\sigma}_n$ is well defined. A similar definition holds for $Y \neq \emptyset$ and the homomorphism σ_n satisfies Property a) of Theorem (4.4) (i.e. σ_n is natural and commutes with boundary homomorphisms. Therefore, by the five lemma, σ_n will be an isomorphism for $n \geq 7$ (or $n = 6$ and Y empty) if and only if this is true when Y is empty. The proof of this fact goes as the proof of Proposition (4.7), using the Cappell-Shaneson surgery theory with coefficients instead of the theory of Wall. We give here the details for the injectivity of $\sigma_n(\hat{j}^X)$ when n is even, to illustrate the use of the hypothesis that $\mathcal{F}$ is locally epic. The other arguments are similar.

As in the proof of (4.7), one may suppose that X is connected and the existence of the lifting $\widetilde{j}^X$ provides a decomposition

$$\Omega_n^{\Lambda P}(X) \cong \Omega_n^{Q\Lambda P}(X) \oplus \Omega_n^M(X)$$

The injectivity of σ_n is then equivalent to $\ker\overline{\sigma}_n \subset \Omega_n^M(X)$. Let (P,f) represent an element of $\Omega_n^P(X)$ whose image under $\overline{\sigma}_n$ is equal to zero. Let

$$\begin{array}{ccc} \nu_M & \longrightarrow & \xi \\ \downarrow & & \downarrow \\ M & \xrightarrow{\ g\ } & P \end{array}$$

be a surgery problem associated to $(\widetilde{j}^X \circ f,(P))$, with surgery obstruction $\sigma(g) \in \Gamma_n(Z\pi_1(P) \to \Lambda)$. By hypothesis, one has $f_*(\sigma(g)) = 0$ in $\Gamma_n(Z\pi_1(X) \to \Lambda)$. By Lemma (8.2.a), there exists a connected space X_0 with finitely presented fundamental group and a factorization

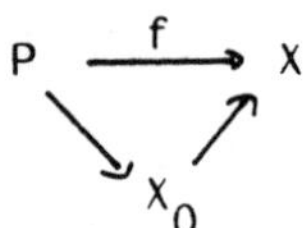

with $(f_0)_*(\sigma(g)) = 0$ in $\Gamma_n(Z\pi_1(X_0) \to \Lambda)$. As $n \geq 4$ and as P has a smooth 2-skeleton, , one may perform low dimensional surgeries on P and M (using (3.17)) in order to transform (g,f,f_0) into $(g',f'.f_0')$

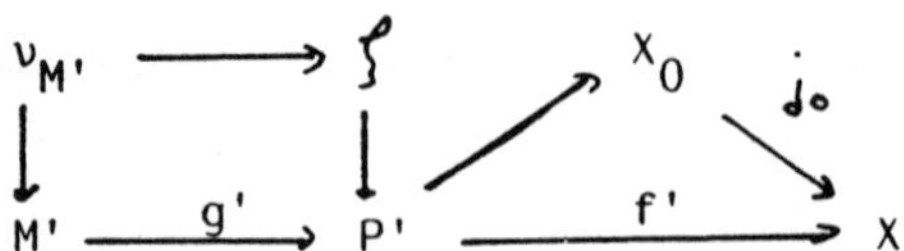

with $(f_0)_*(\sigma(g')) = 0$ and $\pi_1(P') = \pi_1(X_0)$. Then $\sigma(g') = 0$.

Suppose first that $Z\pi_1(P') \to \Lambda$ is locally epic. As $n \geq 6$, there is, by [CS, Theorem 1.7], a cobordism $(W,M'M'')$ and an extension of g' to a normal map $G : W \to P'$ such that $g'' = G$ $M'' : M'' \to P'$ is a Λ-homology equivalence inducing an isomorphism on fundamental groups. The space $R = W \cup \mathcal{M}_{g''}$ is a Poincaré cobordism from (P',f') to $(M'',f' \circ g'')$. The later represents an element of $\Omega_n^M(X)$; this proves that $\ker\bar{\sigma}_n \subset \Omega_n^M(X)$ which, by what was said above, would prove the injectivity of σ_n for n even ≥ 6.

In general, we may not suppose that $Z\pi_1(P') \to \Lambda$ is locally epic. One can still make g' (k-1)-connected (n = 2k). The hypothesis locally epic is then used to kill the last kernel $K_k(M') \otimes \Lambda$ of g' (see [CS, p.293]). This requires as usual to realize a prefered basis of a free Λ-module $H \otimes \Lambda$ (a subkernel) by of the $Z\pi_1(P')$-module H. By Lemma (8.2.b), there exists a finitely presented group T and a factorization $\pi_1(P') \to T$ $\pi_1(X)$ such the basis of $H \otimes \Lambda$ can be realized by elements of $H \otimes ZT$.

As P' has a smooth 2-skeleton, one can, by performing 0 and 1 dimensional surgeries on M' and P', obtain a normal cobordism between g' and $g'' : M'' \to P''$, so that :

- g'' is (k-1)-connected
- $\pi_1(P'') = T$
- $K_k(M'') = K_k(M') \otimes Z\pi_1(T)$

(Observe that T may be chosen of the form $T = \pi_1(P')*T$ and thus M'' and P" would be obtained by boundary connected sum of M' and P' with some thickening of a 2 dimensional complex). One has now that $\sigma(g'') = 0$ in $\Gamma_n(Z\pi_1(P'') \to \Lambda)$. The last surgeries can then be performed on M'', as in [CS, p. 293], to the Λ-homology equivalence $g'' : M'' \to P''$, as above. □

As in the proof of Theorem (4.4), the rest of the proof of Theorem (8.1) consists of proving the following result :

(8.3) **Lemma** Suppose that $\mathcal{F}$ is locally epic. If $\pi_1(X) \cong \pi_1(Y)$, then $\Omega_n^{Q\Lambda P}(X,Y) = 0$ for $n \geqslant 7$.

Lemma (8.3) will follow from Lemma (8.7) below. We need some preliminaries.

(8.4) **Definition** Let $g : (P,\partial P) \to (R,\partial R)$ be a morphism of Λ-Poincaré spaces. Then g is called a **Λ-equivalence** if the homomorphisms :

$$\pi_1 g : \pi_1(P) \to \pi_1(R),$$
$$\pi_1(g|\partial P) : \pi_1(\partial P) \to \pi_1(\partial R),$$
$$g_* : H_*(P;\Lambda) \to H_*(R;\Lambda)$$
$$(g|\partial P)_* : H_*(\partial P;\Lambda) \to H_*(\partial R;\Lambda)$$

are all isomorphisms.

(8.5) **Lemma** Let $f : R \longrightarrow X$ be a map, with R a Λ-Poincaré space of formal dimension n, satisfying $\pi_1(\partial R) \cong \pi_1(R)$. Suppose that $Z\pi_1(R) \longrightarrow \Lambda$ is locally epic. Then, if $n \geqslant 6$, there exists a Λ-equivalence $g : (P,\partial P) \longrightarrow (R,\partial R)$, where P is a Poincaré space.

Proof : As $\pi_1(\partial R) \cong \pi_1(R)$, one has $\Omega_n^P(R,\partial R) = \Omega_n^Q(R,\partial R)$ for all n by Theorem (4.4). Therefore, the class $[id_R]$ has the same image in $\Omega_n^Q(R,\partial R)$ that a class $[g : (P,\partial P) \longrightarrow (R,\partial R)]$ in $\Omega_n^P(R;\partial R)$. Then g is a morphism of Λ-Poincaré spaces with P a Poincaré space. As $\pi_1(\partial R) \cong \pi_1(R)$, and if $Z\pi_1(R) \longrightarrow \Lambda$ is locally epic, the classical proof of the $(\pi\text{-}\pi)$-theorem [Wa2, Chapter 4] may be worked out when $n \geqslant 6$, using the techniques of Section 5.C. This gives a Poincaré bordism between g and a Λ-homology equivalence $g : (P,\partial P) \longrightarrow (R,\partial R)$. □

(8.6) **Lemma** Let $f : (R,\partial R) \longrightarrow (X,Y)$ be a map, with R a Λ-Poincaré space of formal dimension $n \geqslant 6$. Suppose that $\pi_1(X) \cong \pi_1(Y)$ and that $Z\pi_1(X) \longrightarrow \Lambda$ is locally epic. Then f is Λ-Poincaré bordant to $f' : (P,\partial P) \longrightarrow (X,Y)$, where P is a Poincaré space.

Proof : As Λ-Poincaré spaces have a smooth 2-skeleton, and as $\pi_1(Y) \cong \pi_1(X)$, one can perform surgeries on R and ∂R such that $\pi_1(\partial R) \cong \pi_1(R)$. One can then apply the proof of (8.5). The hypothesis locally epic is used at the last step of the surgery, as in the proof of (8.1). When only $Z\pi_1(X) \to \Lambda$ is locally epic, one adapts the fundamental group of R, as in the proof of (8.1), by making boundary connected sums with some thickening of a 2-complex. []

(8.7) **Lemma** If $\pi_1(Y) \cong \pi_1(X)$ is an isomorphism, then the homomorphisms

$$\Omega_n^P(X,Y) \longrightarrow \Omega_n^{\Lambda P}(X,Y) \longrightarrow \Omega_n^Q(X,Y)$$

are isomorphisms for $n \geq 6$.

Observe that Lemma (8.7) implies Lemma (8.3), which terminates the proof of Theorem (8.1).

Proof of (8.7) : Because of the commutativity of the following diagram :

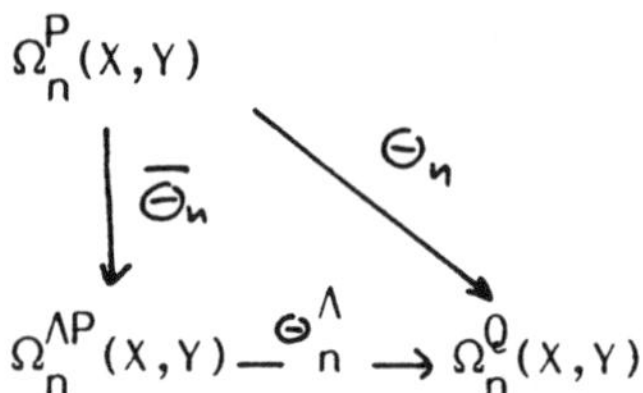

and the fact that Θ_n is an isomorphism when $\pi_1(X) \cong \pi_1(Y)$ (by Theorem (4.4)), then, for all n, the homomorphism Θ_n^Λ is surjective and $\overline{\Theta}_n$ is injective. It then suffices to prove that $\overline{\Theta}_n$ is onto for $n \geq 6$. This follows from Lemma (8.6). []

We now give some additional facts about Λ-Poincaré bordism groups as well as consequences of our previous lemmas.

(8.8) **Lemma** Let us consider $\Lambda = Z\pi_1(X)$, with the homomorphism $Z\pi_1(X) \longrightarrow \Lambda$ being the identity. Then, the natural homomorphism $\Omega_n^P(X) \longrightarrow \Omega_n^{\Lambda P}(X)$ is an isomorphism for all $n \geqslant 4$ and is injective for $n = 3$.

Proof : Let $f : R \longrightarrow X$ represent an elememt of $\Omega_n^{\Lambda P}(X)$. If $n \geqslant 4$, surgery of index ≤ 1 can be made on R, using Lemma (8.6). Hence, f is Λ-Poincaré bordant to $g : P \longrightarrow X$ so that $\pi_1 g$ is injective. Let B be a $\pi_1(P)$-module. As P is a $Z\pi_1(X)$-Poincaré space, one has :

$$\begin{array}{ccccc} H^i(P;B) & \longrightarrow & H^i(P;B \otimes_{Z\pi_1(P)} Z\pi_1(X)) & \cong & \oplus H^i(P;B) \\ \downarrow \cap[P] & & \downarrow \cap[P] & & \downarrow (\oplus \cap[P]) \\ H_{n-i}(P;B) & \longrightarrow & H_{n-i}(P;B \otimes_{Z\pi_1(P)} Z\pi_1(X)) & \cong & \oplus H_{n-i}(P;B) \end{array}$$

where the direct sum is taken over the set of left cosets $\pi_1(X)/\pi_1(P)$. The composed horizontal homomorphisms are the inclusion onto the summand corresponding to the coset of 1. Therefore the cap product with [P] is an isomorphism $H^i(P;B)$ $H_{n-i}(P;B)$ for any $\pi_1(P)$-module B and then P is a Poincaré space. This shows the surjectivity of $\Omega_n^P(X) \longrightarrow \Omega_n^{\Lambda P}(X)$ when $n \geqslant 4$. The same procedure applied to Λ-Poincaré cobordisms gives the injectivity of $\Omega_n^P(X) \longrightarrow \Omega_n^{\Lambda P}(X)$ when $n \geqslant 3$. []

We suspect Theorem (8.1) to be true when $n \geq 6$ (or $n = 5$ and $Y = \emptyset$), but our techniques do not enable us to establish it (because of the dimensional restriction in Proposition (8.3)). There is however a special case where it is possible, which was used in [HH]. Recall that a group G is **locally perfect** if every finitely generated subgroup of G is contained in a finitely generated perfect subgroup of G.

(8.9) <u>**Proposition**</u> Let (X,Y) be a pair over BG. Let P be a locally perfect normal subgroup of $\pi_1(X)$ and let $\Lambda = Z[\pi_1(X)/P]$. Then :

a) $\Omega_n^{Q\Lambda P}(X,Y) = 0$ for $n \geq 5$, when $\pi_1(Y) = \pi_1(X)$

b) If $n \geq 5$ and $Y = \emptyset$, there is a natural isomorphism

$$\sigma_n : \Omega_{n+1}^{Q\Lambda P}(X) \longrightarrow \Gamma_n(Z\pi_1(X) \rightarrow \Lambda) = L_n(\pi_1(X)/P;w^X)$$

<u>**Remark**</u> : The above-mentioned isomorphism $\Gamma_n(Z\pi_1(X) \longrightarrow \Lambda) = L_n(\pi_1(X)/P;w^X)$ can be easily deduced from [Ha1] (the case where $\pi_1(X)$ is finitely generated is done) or more directely from [Vo, Theorem (9.7)].

The proof of Proposition (8.9) will be given after that of Lemma (8.10) below. Let X, P and Λ as in Proposition (8.9) (case $Y = \emptyset$). There is an esentially unique acyclic map (i.e. a map whose theoretical fiber is acyclic; in another terminology, a Quillen plus construction) $i : X \longrightarrow X^+$ so that

$\pi_1(X^+) = \pi_1(X)/P$ (See [HHu]). As $P \subset \ker \pi_1 j^X$, there are factorizations $X^+ \to BG$ of j^X, all in the same homotopy class [HHu, Proposition (3.1)], and hence X^+ can be considered as a space over BG by chosing one of these.

(8.10) **Lemma** The above acyclic map $i : X \to X^+$ induces, for all n, an isomorphism :

$$i_* : \Omega_n^{\Lambda P}(X) \xrightarrow{\simeq} \Omega_n^{\Lambda P}(X^+).$$

Proof : Let $f : R \to X$ represent an element of $\Omega_n^{\Lambda P}(X^+)$. Form the pull-back diagram :

$$\begin{array}{ccc} \bar{R} & \xrightarrow{\bar{f}} & X \\ \downarrow & & \downarrow i \\ R & \xrightarrow{f} & X^+ \end{array}$$

The map $\bar{R} \to R$ is an acyclic map and $\ker(\pi_1(\bar{R}) \to \pi_1(R))$ is locally perfect. Moreover, R is homotopy equivalent to finite complex, being a Λ-Poincaré space. By [HV, Theorem 3.1], there is a finite complex R and a commutative diagram

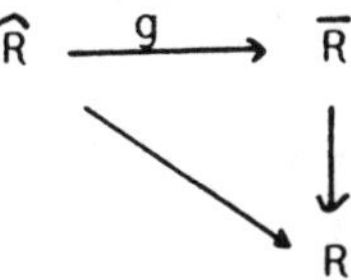

where $\hat{R} \to R$ is an acyclic map. Therefore, $\hat{R}$ is a Λ-Poincaré space and $\bar{f} \circ g$ represents an element of $\Omega_n^{\Lambda P}(X)$. The mapping cylinder of $\hat{R} \to R$ constitutes a Λ-Poincaré cobordism between

$i \circ \bar{f} \circ g$ and f. This shows the surjectivity of i_*. The same argument applied to Λ-Poincaré bordisms relative to their ends gives the injectivity of i_*. []

Proof of Proposition (8.9) : As $Z\pi_1(X) \longrightarrow \Lambda$ is locally epic in our case, one can follow the procedure of the proof of Theorem (8.1). It is then enough to prove Statement a), which improves Lemma (8.3) in our special case. As $\pi_1(Y) \cong \pi_1(X)$, P is also a normal subgroup of $\pi_1(Y)$ and the acyclic map $Y \longrightarrow Y^+$ is defined. Since i induces an isomorphism on the integral homology, it induces isomorphisms $\Omega_*^Q(X) \xrightarrow{\simeq} \Omega_*^Q(X^+)$, $\Omega_*^Q(Y) \xrightarrow{\simeq} \Omega_*^Q(Y^+)$ and $\Omega_*^Q(X,Y) \xrightarrow{\simeq} \Omega_*^Q(X^+,Y^+)$ (using, for instance, Spectral Sequence (3.11)). Therefore, using the long exact sequences involving the bordism groups $\Omega_*^{\Lambda P}$, Ω_*^Q and $\Omega_*^{Q\Lambda P}$ for the pairs (X,Y) and (X^+,Y^+), Lemma (8.10) and the five lemma, one obtains the isomorphism :

$$\Omega_*^{Q\Lambda P}(X,Y) \xrightarrow{\simeq} \Omega_*^{Q\Lambda P}(X^+,Y^+).$$

On the other hand, if one uses the following diagram :

$$\begin{array}{ccccccccc}
\Omega_n^P(Y^+) & \longrightarrow & \Omega_n^P(X^+) & \longrightarrow & \Omega_n^P(X^+,Y^+) & \longrightarrow & \Omega_{n-1}^P(Y^+) & \longrightarrow & \Omega_{n-1}^P(X^+) \\
\downarrow & & \downarrow & & \downarrow & & \downarrow & & \downarrow \\
\Omega_n^{\Lambda P}(Y^+) & \longrightarrow & \Omega_n^{\Lambda P}(X^+) & \longrightarrow & \Omega_n^{\Lambda P}(X^+,Y^+) & \longrightarrow & \Omega_{n-1}^{\Lambda P}(Y^+) & \longrightarrow & \Omega_{n-1}^{\Lambda P}(X^+)
\end{array}$$

and Lemma (8.8), one obtains the isomorphism :

$\Omega_n^P(X^+,Y^+) \xrightarrow{\simeq} \Omega_n^{\Lambda P}(X^+,Y^+)$ for $n \geqslant 5$. Part a) of Proposition (8.9) then follows from Theorem (4.4). []

A consequence of Theorem (8.1) is the formal result of surgery for Λ-Poincaré spaces, in the spirit of Section 5.A. Let P be a Λ-Poincaré space of formal dimension n with a Λ-Poincaré decomposition $\partial P = \partial_- P \cup \partial_+ P$ of ∂P. Let $\mathcal{T}^\Lambda(P,\partial_+P)$ be the set of Λ-Poincaré bordism classes of morphisms of Λ-Poincaré spaces :

$$f : (\bar{P},\partial_-\bar{P},\partial_+\bar{P}) \longrightarrow (P,\partial_-P,\partial_+P)$$

where $\partial\bar{P} = \partial_-\bar{P} \cup \partial_+\bar{P}$ is a Λ-Poincaré decomposition of $\partial\bar{P}$ and $f|\partial_-\bar{P} : \partial_-\bar{P} \longrightarrow \partial_-P$ is a Λ-equivalence. Using Theorems (8.1) and Proposition (8.9) instead of Theorem (4.4), one can prove, as in Theorem (5.1) Part 1) :

(8.11) <u>**Theorem**</u> Suppose that $n \geqslant 7$, or $n = 6$ and $\partial P = \emptyset$ and that $Z\pi_1(P) \to \Lambda$ is locally epic. Then, here exists a map :

$$s^\Lambda : \mathcal{T}^\Lambda(P,\partial_+P) \longrightarrow \Gamma_n(\mathfrak{F}), \quad \mathfrak{F} = \begin{pmatrix} Z\pi_1(\partial_+P) & \longrightarrow & \Lambda \\ \downarrow & & \downarrow \\ Z\pi_1(P) & \longrightarrow & \Lambda \end{pmatrix}$$

such that $s^\Lambda(\alpha) = 0$ if and only if α contains a representative $\hat{f} : \hat{P} \longrightarrow P$ such that $\hat{f}$ and $f|\partial_+\hat{P}$ are Λ-equivalences. If (P,∂P) and Λ are as in Proposition (8.9), then the dimension conditions can be weakened to be $n \geqslant 5$. []

In the same spirit, the other results of Chapter 5, 6 and 7.A are available, with convenient local epicity and low-dimensional restrictions. However, transversality up to homotopy equivalence (as in Section 7.B) is a priori not available, because of the use of low-dimensional arguments in induction processes.

We now finish this section by two consequences of our previous results.

(8.12) **Proposition** Let P be a connected Λ-Poincaré space of formal dimension $n \geqslant 6$, with $\partial P = \emptyset$ (to simplify). Suppose that $Z\pi_1(P) \longrightarrow \Lambda$ is locally epic. Then, P determines an element

$$s(P) \in S(P) = \Gamma_n \begin{pmatrix} Z\pi_1(P) & = & Z\pi_1(P) \\ \| & & \downarrow \\ Z\pi_1(P) & \longrightarrow & \Lambda \end{pmatrix}$$

which vanishes if and only if P is Λ-Poincaré cobordant to a Poincaré space.

Proof : The identity of P determines an element $[id_P]$ in $\Omega_n^{\Lambda PP}(P)$ which vanishes if and only if P is Λ-Poincaré cobordant to a Poincaré space. It is then enough to establish an isomorphism α between $\Omega_n^{\Lambda PP}(P)$ and $S(P)$.

Let $f : R \longrightarrow P$ represent a class in $\Omega_n^{\Lambda PP}(P)$. Suppose that j^P admits a lifting $\tilde{j}^P : P \longrightarrow BO$. By transversality, this gives a normal map of degree one $g : V \longrightarrow R$, with V a compact manifold, whose surgery obstruction can be defined in

S(R), by [CS1]. Its image in S(P) will be by definition $\alpha([f];j^P)$. As for Proposition (4.7), one checks that $\alpha(\ ;j^P)$ is well defined and is an homomorphism. To define α in general, we proceed as for Theorem (4.4), by showing that $\Omega_n^{\Lambda PP}(X,Y) = 0$ when the pair (X,Y) is 2-conneted and $n \geq 6$. This comes from the exact sequence :

$$\Omega_{n+1}^{Q\Lambda P}(X,Y) \longrightarrow \Omega_n^{\Lambda PP}(X,Y) \longrightarrow \Omega_n^{QP}(X,Y)$$

together with Propositions (4.8) and Lemma (8.3). Therefore α is defined. Observe that there is the long exact sequence :

$$\Gamma_n(Z\pi_1(P) \to Z\pi_1(P)) \longrightarrow \Gamma_n(Z\pi_1(P) \to \Lambda) \longrightarrow S(P) \longrightarrow$$

$$\longrightarrow \Gamma_{n-1}(Z\pi_1(P) = Z\pi_1(P)) \longrightarrow \Gamma_{n-1}(Z\pi_1(P) \to \Lambda)$$

and that $\Gamma_n(Z\pi_1(P) = Z\pi_1(P)) = L_n(\pi_1(P))$. Commutative diagram (8.12.a) below shows that α is an isomorphism for $n \geq 7$, by Theorems (4.4) and (8.1). []

(8.13) **Proposition** Let R be a Λ-Poincaré space of formal dimension $n \geq 6$ (with $\partial R = \emptyset$, to simplify). Suppose that $Z\pi_1(R) \to \Lambda$ is locally epic. Then, R is Λ-equivalent to $P \cup_h V$, where V is a manifold, thickening of a 2-complex with $\pi_1(V) \cong \pi_1(R)$, P is a Poincaré space and $h : \partial P \to \partial V$ is a Λ-equivalence.

Proof : Let V be a smooth 2-skeleton for R (which exists by (2.15)). One has a Λ-Poincaré decomposition $R = V \cup P_0$,

$V \cap P_0 = \partial V = \partial P_0$, and P_0 is a Λ-Poincaré space, with $\pi_1(\partial P_0) = \pi_1(P_0)$. By (8.5), there is a Λ-equivalence $f : (P,\partial P) \longrightarrow (P_0,\partial P_0)$, where P is a Poincaré space. []

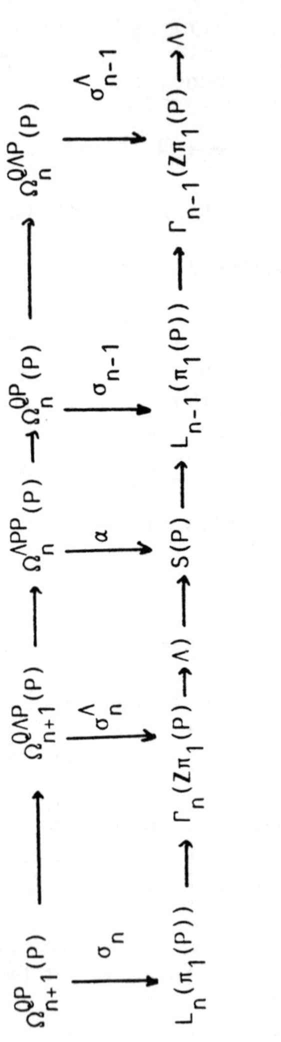

Diagram (8.12.a)

C. NON-FINITE POINCARÉ SPACES

Except in Propositions (2.21) and (3.20), a Poincaré space P has always been supposed to be homotopy equivalent to a finite complex. In this section, we relax this general condition by just requiring that P be dominated by a finite complex. Observe that by [Br2, Corollary 2] or [Bn, Section 3], this is equivalent to $\pi_1(P)$ being finitely presented. There is no known example of a space satisfying Poincaré duality with a non-finitely presented fundamental group. For discussion about this subject, see [Ha2, Section 3].

With this generalized notion of Poincaré spaces, one obtains groups $\Omega_n^{\overline{P}}(X,Y)$, $\Omega_n^{Q\overline{P}}(X,Y)$, etc. The result corresponding to Theorem (4.4) uses the projective Wall obstruction group L_n^p and is the following :

(8.14) <u>Theorem</u> : For $n \geqslant 6$, or $n \geqslant 5$ and $Y = \emptyset$, there is a natural isomorphism :

$$\sigma_n^p : \Omega_{n+1}^{Q\overline{P}}(X,Y) \longrightarrow L_n^p(\pi_1(X),\pi_1(Y);\pi_1 j^X)$$

which commutes with the boundary homorphisms.

<u>Proof</u> : If j^X admits a lifting $\tilde{j}^X$ through BO, then a homorphism $\sigma_n^p(\tilde{j}^X) : \Omega_{n+1}^{Q\overline{P}}(X,Y) \longrightarrow L_n^p(\pi_1(X),\pi_1(Y);\pi_1 j^X)$ is defined for $n \geqslant 5$, as in Proposition (4.7) which is natural and commutes with boundary homomorphisms. Therefore, by the

five lemma, $\sigma_n(\tilde{j}^X)$ satisfies the conclusion of (8.14) if and only if this is the case for Y empty. The proof of this fact goes as the proof of Proposition (4.7), using the projective surgery techniques (essentially [PR, theorem 2.1]) instead of the surgery for finite Poincaré spaces. The low dimensional surgery arguments which are necesary in places to adapt fundamental groups can be done, using Proposition (3.20).

As in the proof of Theorem (4.4), the rest of the proof of Theorem (8.14) consists of establishing the following result :

(8.15) **Lemma** If (X,Y) is 2-connected, then $\Omega_n^{Q\bar{P}}(X,Y) = 0$ for $n \geq 6$.

Proof : Let us consider the exact sequence :

$$\Omega_n^{QP}(X,Y) \longrightarrow \Omega_n^{Q\bar{P}}(X,Y) \longrightarrow \Omega_{n-1}^{\bar{P}P}(X,Y).$$

Then, Proposition (8.15) will follow from Proposition (4.8) together with the fact that $\Omega_n^{\bar{P}P}(X,Y) = 0$ when (X,Y) is 2-connected and $n \geq 5$. The later comes from the following exact sequence :

$$\Omega_{n+1}^{\bar{P}P}(X,Y) \longrightarrow \Omega_n^{\bar{P}P}(Y) \longrightarrow \Omega_n^{\bar{P}P}(X) \longrightarrow \Omega_n^{\bar{P}P}(X,Y)$$

together with Lemma (8.16) below. []

(8.16) **Lemma** There is a homomorphism $\alpha_n : \Omega_n^{\bar{P}P}(X) \longrightarrow H^n(Z_2;\tilde{K}_0(Z\pi_1(X)))$ which is bijective when $n \geq 5$ and injective for $n = 4$.

<u>Proof</u> : Let $f : P \longrightarrow X$ represent an element $a \in \Omega_n^{\bar{P}P}(X)$. Let $\alpha(P) \in \tilde{K}_0(Z\pi_1(P))$ be the Wall finiteness obstruction of P. Classical Poincaré duality arguments show that $\alpha(P) + (-1)^{n+1}\alpha(P)^*$ is equal to the image in $\tilde{K}_0(Z\pi_1(P))$ of $\alpha(\partial P)$. As ∂P is a finite complex by hypothesis, one has $\alpha(P) + (-1)^{n+1}\alpha(P)^* = 0$ and then $\alpha(P)$ determines an element of $H^n(Z_2;\tilde{K}_0(Z\pi_1(P)))$. Its image in $H^n(Z_2;\tilde{K}_0(Z\pi_1(X)))$ under the homomorphism induced by f will be by definition $\alpha_n(a)$. If $g : W \longrightarrow X$ is a $(\bar{P}P)$-bordism between f and $f' : P' \longrightarrow X$, then the equation $\alpha(P') = \alpha(P) + \alpha(W) + (-1)^n\alpha(W)^*$ holds in $\tilde{K}_0(Z\pi_1(W))$, which shows that α_n is a well defined homomorphism from $\Omega_n^{\bar{P}P}(X)$ to $H^n(Z_2;\tilde{K}_0(Z\pi_1(X)))$.

We first give an argument showing that α_n is bijective for $n \geq 6$ and injective for $n \geq 5$. The idea of this argument was given to us by A. Ranicki.

Let $b \in H^n(Z_2;\tilde{K}_0(Z\pi_1(X)))$. The group $H^n(Z_2;\tilde{K}_0(Z\pi_1(X)))$ is the inductive limit of the groups $H^n(Z_2;\tilde{K}_0(ZG))$, for the inductive system of homomorphisms $G \longrightarrow \pi_1(X)$, with G finitely presented. Therefore, there exists a finitely presented group G with a homomorphism $G \longrightarrow \pi_1(X)$, such that there is an element $c \in H^n(Z_2;\tilde{K}_0(ZG))$ having image b in $H^n(Z_2;\tilde{K}_0(Z\pi_1(X)))$. The class c is represented by $d \in \tilde{K}_0(ZG)$ satisfying $d = (-1)^n d^*$.

By the classical realization result of Wall [Wa3], there is a finitely dominated complex K with $\pi_1(K) = G$, with finiteness obstruction $\alpha(K) = d$. The complex K can be chosen of cohomology dimension 2. For such a K, there is no obstruction to constructing a map $K \longrightarrow X$ realizing the homomorphism $G \longrightarrow \pi_1(X)$.

The complex $K \times S^1$ is of cohomology dimension 3 and has zero Wall finiteness obstruction. Therefore, there is a homotopy equivalence $Z \xrightarrow{\simeq} K \times S^1$, with Z a finte 3-dimensional complex. Let $r : Z \longrightarrow X$ be the composed map

$$Z \longrightarrow K \times S^1 \longrightarrow K \longrightarrow X.$$

We consider Z as a space over BG by $j^Z = j^X \circ r$. As j^Z factors through a cohomology dimension 2 complex, there exists a lifting $\tilde{j}^Z : Z \longrightarrow BO$ of j^Z. By [Wa1, pp.79-80], there is, for $n \geq 6$, a thickening V^n of Z (i.e. V collapses onto Z and $\pi_1(\partial V) = \pi_1(V)$), with stable normal bundle induced by $\tilde{j}^Z$. Let $(\overline{V}, \overline{\partial V})$ be the infinite cyclic covering of $(V, \partial V)$.

The open manifold $\overline{V}$ is homotopy equivalent to K. We now use the argument of [Ra, preliminaries of Proposition (3.5)] which establishes that $\overline{\partial V}$ is finitely dominated. For the convenience of the reader, we recopy the proof here : "Thicken up the self-homotopy equivalence transposing the S^1-factors

$$\mathrm{id} \times T : K \times S^1 \times S^1 \longrightarrow K \times S^1 \times S^1 \;;\quad (x,s,t) \longmapsto (x,t,s)$$

to a self homotopy equivalence of a pair

$$(g,\partial g) \ : \ (V,\partial V) \times S^1 \longrightarrow (V,\partial V) \times S^1$$

inducing on the fundamental group the automorphism

$$K \times \mathbb{Z} \times \mathbb{Z} \longrightarrow K \times \mathbb{Z} \times \mathbb{Z} \ ; \ (x,s,t) \longmapsto (x,t,s)$$

transposing the $\mathbb{Z}$-factors. Thus $(g,\partial g)$ lifts to a $\mathbb{Z}$-equivariant homotopy equivalence

$$(\overline{g},\overline{\partial g}) \ : \ (\overline{V},\overline{\partial V}) \times S^1 \longrightarrow (V,\partial V) \times \mathbf{R} \ .$$

In particular, this shows that ∂V is a finite complex with a finitely dominated infinite cyclic cover $\overline{\partial V}$.

It has long been noticed that the fact that $\overline{\partial V}$ is finitely dominated implies that $\overline{\partial V}$ satisfies Poincaré duality (the current most general result in this context is to be found in [Ba]). Poincaré Duality classical arguments show that $\alpha(\overline{\partial V}) = \alpha(V) + (-1)^{n+1}\alpha(V)^* = d + (-1)^{n+1}d^* = 0$. Therefore, $\overline{\partial V}$ is a Poincaré space and the map :

$$\overline{V} \longrightarrow V \longrightarrow Z \xrightarrow{\ r\ } X$$

represents a class $e \in \Omega_n^{\overline{P}P}(X)$, with $\alpha_n(e) = b$ in $H^n(\mathbb{Z}_2;\widetilde{K}_0(\mathbb{Z}\pi_1(X)))$. This shows the surjectivity of α_n when $n \geq 6$.

To show the injectivity of α_n for $n \geq 5$, let $\Sigma : P \longrightarrow X$ representing $e \in \Omega_n^{\overline{P}P}(X)$ such that $\alpha_n(e) = 0$. The group $H^n(\mathbb{Z}_2;\widetilde{K}_0(\mathbb{Z}\pi_1(X)))$ being the inductive limit of the groups $H^n(\mathbb{Z}_2;\widetilde{K}_0(\mathbb{Z}G))$, for the inductive system of homomorphisms

$G \longrightarrow \pi_1(X)$, with G finitely presented, there exists a finitely presented group G with homomorphisms $\pi_1(P) \to G \to \pi_1(X)$, such that $\alpha(P)$ has image zero in $H^n(Z_2;\widetilde{K}_0(ZG))$. By low dimensional surgeries, using Proposition (3.20), one may suppose that $\pi_1(\partial P) \cong \pi_1(P) \cong G$, and therefore, $\alpha(P) = \beta + (-1)^n\beta^*$, for some $\beta \in \widetilde{K}_0(Z\pi_1(P))$. The above construction permits us to obtain a finitely dominated poincaré pair $(\overline{V},\overline{\partial V})$ of formal dimension n+1, a map $s : \overline{V} \longrightarrow P$ inducing an isomorphism on the fundamental groups, with $\alpha(\overline{V}) = -\beta$. Then, $\alpha(\overline{\partial V}) = -\alpha(P)$.

Observe that $f \circ s : \overline{\partial V} \longrightarrow X$ represents the zero element in $\Omega_n^{\overline{P}P}(X)$ (even in $\Omega_n^{\overline{P}}(X)$). Therefore, the element e in $\Omega_n^{\overline{P}P}(X)$ can be represented by $\hat{f} : \hat{P} \longrightarrow X$, where $\hat{P}$ is obtained by taking the connected sum of P with $\overline{\partial V}$ and then performing low dimensional surgeries to have $\pi_1(\hat{P}) \cong \pi_1(P)$. But then $\alpha(\hat{P}) = 0$, which implies that $e = 0$.

To obtain the information in lower dimensions, let us consider the pull-back diagram

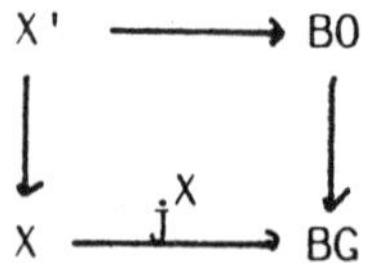

and the diagram

$$\begin{array}{ccc}\Omega_n^{\bar{P}P}(X') & \xrightarrow{\alpha'_n} & H^n(Z_2;\tilde{K}_0(Z\pi_1(X'))) \\ \downarrow & & \downarrow \\ \Omega_n^{\bar{P}P}(X) & \xrightarrow{\alpha_n} & H^n(Z_2;\tilde{K}_0(Z\pi_1(X)))\end{array}$$

The techniques of [PR,p.244] actually show that α'_n is bijective for $n \geqslant 5$ and injective for $n = 4$. Therefore, α_n is surjective for $n \geqslant 5$. (This gives a second proof of the surjectivity of α_n for $n \geqslant 6$; but we could not deduce the injectivity of α_n from the above diagram for large n).

It remains to prove the injectivity of α_4. By what was said above it is enough to show that the homomorphism $\Omega_n^{\bar{P}P}(X') \longrightarrow \Omega_n^{PP}(X)$ is surjective for $n = 4$. In fact, this is true for $n \leq 5$. Indeed, let $f : P \longrightarrow X$ represent an element of $\Omega_n^{\bar{P}P}(X)$. If $n \leq 5$, one can subtract handles of $(P,\partial P)$, using (3.20), so that the pair $(P,\partial P)$ be $(n-3)$-connected. Therefore, as in the proof of Lemma (4.13), one proves that there is no obstruction to obtain a lifting $f : P \longrightarrow X'$ of f. []

9. OTHER APPLICATIONS

This section contains the proof of a few facts announced in [Qu] which are not avaible in the literature.

Let (X,A) and (Y,B) be pairs over BG. The space $X\times Y$ is considered over BG by taking for $j^{X\times Y}$ the composition of $j^X\times j^Y : X\times Y \to BG\times BG$ with the map $BG\times BG \to BG$ given by the Whitney sum. If $f : (Q,\partial Q) \to (X,A)$ and $f' : (Q',\partial Q') \to (Y,B)$ represent elements of $\Omega_n^Q(X,A)$ and $\Omega_p^Q(Y,B)$, then $f\times f'$ represent an element of $\Omega_{n+p}^Q(X\times Y, A\times Y \cup X\times B)$. This procedure gives rise to bilinear maps

$$\Omega_n^Q(X,A) \times \Omega_p^Q(Y,B) \longrightarrow \Omega_{n+p}^Q(X\times Y, A\times Y \cup X\times B)$$

$$\Omega_n^P(X,A) \times \Omega_p^Q(Y,B) \longrightarrow \Omega_{n+p}^Q(X\times Y, A\times Y \cup X\times B)$$

$$\mu : \Omega_n^P(X,A) \times \Omega_p^{QP}(Y,B) \longrightarrow \Omega_{n+p}^{QP}(X\times Y, A\times Y \cup X\times B).$$

Using the homomorphism σ_* of Theorem (4.4), the last map μ defines, (provisionnaly for $n \geq 5$), a bilinear map :

$$\bar{\mu} : \Omega_n^P(X,A) \times L_m(\pi(Y),\pi(B);j^Y) \longrightarrow L_{m+n}(X\times Y, A\times Y \cup X\times B; j^Y)$$

such that $\sigma_{m+n}(\mu(a,u)) = \mu(a,\sigma_m(u))$. Let P be a closed oriented PD-space of formal dimension n. It determines an element $[P] \in {}^{or}\Omega_n^{(PD)}(pt) = \Omega_n^P(BSG)$ and thus induces a linear map

$$\bar{\mu}([P],-) : L_m(\pi(Y),\pi(B);j^Y) \longrightarrow L_{m+n}(\pi(Y),\pi(B);j^Y).$$

(9.1) **Proposition** The map $\mu([CP^2],-)$ is the periodicity map.

Proof : Without loss of generality, one can replace Y and B by their 2-dimensional skeleta, and then assume that j^Y admits a lifting $\tilde{j}^Y$ through BO. Let $a \in L_m(\pi(Y),\pi(B);j^Y)$. By Proposition (4.9) and the proof of (4.4), there exists

.- a map $g : (R,\partial R) \to (Y,B)$, where R is a Poincaré space of formal dimension m

.- a normal map of degree one $f : (M,\partial M) \to (R,\partial R)$ (M a smooth manifold) with surgery obstruction $\sigma(f) \in L_m(\pi(R),\pi(\partial R);j^R)$; the map from the mapping-cylinder $\mathcal{M}(f)$ to Y extending g represents a class $\alpha \in \Omega^{QP}_{m+1}(Y,B)$

so that $a = \sigma_m(\alpha) = g_*(\sigma(f))$. Therefore, $\mu([CP^2],a)$ is the image by g_* of the surgery obstruction of the map $id \times f : C^2P \times M \to CP^2 \times R$, which is the same as the surgery obstruction of f, via the periodicity isomorphism (see [Wa, p. 265]).[]

Remark : The map μ obviously commutes with the multiplication with CP^2. This permits us to define the bilinear map μ for all integers m.

(9.2) **Proposition** Let (X,A) be a pair over BG. Let $a \in \ker[\Omega^{QP}_{n+1}(X,A) \to \Omega^{P}_{n}(X,A)]$. Then, $8 \cdot a = 0$.

The proof of (9.2) uses the following lemma :

(9.3) **Lemma** : The signature map $sg : {}^{or}\Omega_4^{(PD)}(pt) \to \mathbb{Z}$ is an isomorphism.

Proof : As $sg(CP^2) = 1$, the signature map is surjective. Let A_1 and A_2 be two classes in ${}^{or}\Omega_4^{(PD)}(pt)$ with $sg(A_1) = sg(A_2)$. By (2.16), a PD-space of formal dimension 4 admits a 1-smooth skeleton. Therefore, 1-dimensional surgery is possible and the class A_i admits a 1-connected representative P_i. The signature map provides an isomorphism between the Witt group of $\mathbb{Z}$ and the integers [MH, Corollary 4.4]. Hence, ba making connected sums of P_i with copies of $S^2 \times S^2$, one may find representatives P_1' and P_2' of A_1 and A_2 having isometric intersection forms. By a result of J.H.C. Whitehead ([Wh2], see also [MH, p. 103]), the spaces P_1' and P_2' have the same oriented homotopy type. Therefore $A_1 = A_2$. []

Proof of (9.2) : By Theorem (4.4), $2\cdot\Omega_{n+1}^{QP}(X,A) = 0$ if $n \leq 2$. Thus, an argument is needed only if $n \geqslant 3$. Let $(Q_0, \partial Q_0)$ be a generator of $\Omega_5^{QP}(pt) = L_4(0) = \mathbb{Z}$. The boundary ∂Q_0 of the Q-space Q_0 is a Poincaré space of formal dimension 4 with signature 8. The following diagram :

$$\begin{array}{ccc}
\Omega_{n+1}^{QP}(X,A) & \xrightarrow{\ \partial\ } & \Omega_n^{P}(X,A) \\
\downarrow{\scriptstyle \mu([\partial Q_0],-)} & & \downarrow{\scriptstyle \mu(-,[Q_0])} \\
\Omega_{n+5}^{QP}(X,A) & \xrightarrow{\ (-1)^n\ } & \Omega_{n+5}^{QP}(X,A)
\end{array}$$

is commutative. Indeed, represent $a \in \Omega_{n+1}^{QP}(X,A)$ by a map $f : Q \longrightarrow X$, where ∂Q is Q-decomposed $\partial Q = \partial_- Q \cup \partial_+ Q$, $\partial_- Q \cap \partial_+ Q = \partial_0 Q$, $\partial_- Q$ is a Poincaré space and $f(\partial_+ Q) \subset A$. Then

.- $\partial(a)$ is represented by $f : (\partial_- Q, \partial_0 Q) \longrightarrow (X,A)$

.- $\mu(\partial(a),[Q_0])$ is represented by $g : (\partial_- Q\times Q_0, \partial_0 Q\times Q_0) \longrightarrow (X,A)\times pt \longrightarrow (X,A)$

.- $\mu([\partial Q_0],a)$ is represented by $g' : (\partial Q_0\times Q, \partial Q_0\times\partial_+ Q) \longrightarrow pt\times(X,A) \longrightarrow (X,A)$

The existence of the map $Q_0\times Q \longrightarrow X$ shows that $[g'] = (-1)^n[g]$. therefore, if $a \in \ker\partial$, then $\mu([\partial Q_0],a) = 0$ and one has

$$0 = \sigma_{n+4}(\mu([\partial Q_0],a)) = \overline{\mu}([\partial Q_0],\sigma_n(a))$$

By (9.3), one has $[\partial Q_0] = 8\cdot[CP^2]$ and, using (9.1), one deduces that $8\cdot\sigma_n(a) = 0$

If $n \neq 3$ or if A is empty, then σ_n is injective by (4.4) and therefore $8\cdot a = 0$. The remaining case is covered by the following lemma

(9.4) **Lemma** Let (X,A) be a pair over BG with A non-empty. Let $a \in \ker[\Omega_4^{QP}(X,A) \longrightarrow \Omega_3^{P}(X,A)]$. Then, $2\cdot a = 0$.

Proof : It is sufficient to give a proof for X connected, in which case $H_0(X,A) = 0$. Let us consider the pull back diagram

$$\begin{array}{ccc} (\overline{X},\overline{A}) & \longrightarrow & BO \\ \downarrow & & \downarrow \\ (X,A) & \longrightarrow & BG \end{array}$$

There is the following commutative diagram

$$\begin{array}{ccccccc}
\Omega_4^P(\bar{X},\bar{A}) & \xrightarrow{\bar{\alpha}} & \Omega_4^Q(\bar{X},\bar{A}) & & & & \\
\downarrow & & \downarrow & & & & \\
\Omega_4^P(X,A) & \xrightarrow{\alpha} & \Omega_4^Q(X,A) & \longrightarrow & \Omega_4^{QP}(X,A) & \xrightarrow{\partial} & \Omega_3^P(X,A)
\end{array}$$

$$\Gamma = \Omega_4^Q\begin{pmatrix} \bar{A} \subset \bar{X} \\ \cap \quad \cap \\ A \subset X \end{pmatrix}$$

By (3.11), $\bar{\alpha}$ is onto and therefore $\ker\partial$ is a subquotient of the group Γ. If $H_0(X,A) = 0$, the Serre spectral sequence of the fibration $G/0 \longrightarrow (X,A) \longrightarrow (X,A)$ shows that :

$$H_*\begin{pmatrix} \bar{A} \subset \bar{X} & \\ \cap \quad \cap & ; Z \\ A \subset X & \end{pmatrix} = \begin{cases} H_1(X,A;Z/2Z) & \text{if } * = 4 \\ & \\ 0 \quad \text{if } * \quad 3. & \end{cases}$$

From the spectral sequence of type (3.13)

$$H_*\begin{pmatrix} \bar{A} \subset \bar{X} & \\ \cap \quad \cap & ; \pi_*^s \\ A \subset X & \end{pmatrix} \Rightarrow \Omega_*^Q\begin{pmatrix} \bar{A} \subset \bar{X} \\ \cap \quad \cap \\ A \subset X \end{pmatrix}$$

one then deduces that $\Gamma = H_1(X,A;Z/2Z)$. Lemma (9.3) follows.
[]

(9.5) **Proposition** Let (X,A) be a pair. Then for all n, the homomorphism

$$e : {}^{or}\Omega_n^{(PD)}(X,A) \longrightarrow H_n(X,A)$$

given by $e(P,f) = f_*([P])$ is surjective.

When X and A are 1-connected, Proposition (9.4)is due to N. Levitt [Le2, Theorem 2.2].

<u>Proof</u> : The result is classical for $n \leq 6$ [Th, Theorem III.3 and Corollary III.7]. We give below an argument which is valid for $n \geq 4$.

Use the exact sequence of (4.4) and (4.5)

$$\Omega^{QP}_{n+1}(X\times BSG, A\times BSG) \xrightarrow{\partial} {}^{or}\Omega^{(PD)}_{n}(X,A) \xrightarrow{\alpha} H_*(X,A;MSG)$$

$$e: {}^{or}\Omega^{(PD)}_{n}(X,A) \to H_*(X,A), \qquad \gamma: H_*(X,A;MSG) \to H_*(X,A)$$

It is known that the map $MSG \longrightarrow K(Z)$ splits as a map of spectra which implies that γ is surjective (see [Le2, p. 208]). Let $a \in H_*(X,A)$. By (9.2), cokerα = ker∂ is killed by 8, and then $8\cdot a$ is in the image of e. On the other hand, by a classical result of Thom, a has an odd multiple representable by a manifold. Therefore $a \in$ Ime. []

APPENDIX

ON FIBRATIONS WITH COMPLEXES HAVING FINITE SKELETA

Let **FS** be the class of connected spaces having the homotopy type of a CW-complex with finite skeleta. The following proposition is probably well known by specialists :

(A.1) **Proposition** Let $F \to E \to B$ be a Serre fibration. If B and F are in **FS** and if Image$(\pi_2(B) \to \pi_1(F))$ is a finitely generated $\pi_1(B)$-module, then $E \in$ **FS**.

Proof : By the homotopy exact sequence of the fibration, the group $\pi_1(E)$ has a presentation with, as generators the union of the generators of $\pi_1(B)$ and of $\pi_1(F)$ and the following four kind of relators :

1) relators of $\pi_1(F)$,
2) av_a, where a is a relator of $\pi_1(B)$ and $v_a \in \pi_1(F)$
3) $aba^{-1}w_{a,b}$, where a is a generator of $\pi_1(B)$, b is a generator of $\pi_1(F)$ and $w_{a,b} \in \pi_1(F)$
4) a set of generators of Image$(\pi_2(B) \to \pi_1(F))$ as a $\pi_1(B)$-module.

Therefore $\pi_1(E)$ is finitely presented. We use now the criterion of K. Brown [Bn, Theorem 4] : A space X is in **FS** if and only if $\pi_1(X)$ is finitely presented and, for any direct system T_j ($i \in J$) of $\pi_1(X)$-modules, the homomorphism

$$\varinjlim H^*(X;T_i) \longrightarrow H^*(X;\varinjlim T_i)$$

is an isomorphism. In our case, setting X = E, there is the following diagram with the Serre spectral sequences :

$$\begin{array}{ccc} \varinjlim H^p(B;H^q(F;T_i)) & \Rightarrow & \varinjlim H^{p+q}(E;T_i) \\ \downarrow & & \\ H^p(B;\varinjlim H^q(F;T_i)) & & \downarrow \\ \downarrow & & \\ H^p(B;H^q(F;\varinjlim T_i)) & \Rightarrow & H^{p+q}(E;\varinjlim T_i) \end{array}$$

By Brown's criterion, the two left hand vertcal arrows are isomorphisms. therefore, the right hand vertical arrow is an isomorphism and, by the Brown's criterion, $E \in$ **FS**. []

We have essentially used Proposition (A.1) in the following application :

(A.2) <u>**Proposition**</u> Let B be a CW-complex in **FS** and let C be a subcomplex of B with $C \in$ **FS**. Suppose that B and C have finite fundamental groups. Let $P \in$ FS and let $f : P \longrightarrow B$ be a Serre fibration. Then $f^{-1}(C) \in$ FS.

<u>Examples of utilization</u> : Proposition (A.2) has been essentially used in Chapter 4 in the following cases :

a) C = G/O or a cone over G/O, $B = \Sigma(G/O)$, P a Poincaré space (Lemma (4.17), (4.20), (4.21)).

b) $B = B_i$, C some subcomplex of B_i with $\pi_1(C) \cong \pi_1(B_i) \cong \{\pm 1\}$, P a Poincaré space (Lemma (4.23)).

<u>Proof of (A.2)</u> : Replace the inclusion map $C \subset B$ by a Serre fibration $C_1 \longrightarrow B$. Replace the map $f^{-1}(C) \quad P \times C$ by a Serre fibration $X \longrightarrow P \times C$. One gets a pull-back diagram of Serre fibrations :

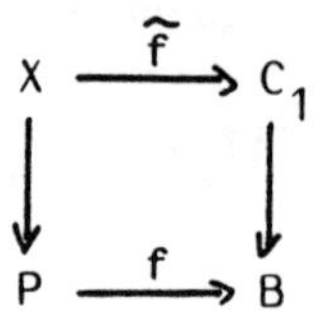

with X homotopy equivalent to $f^{-1}(C)$. Let F be the common fiber $C \to B$ and $X \longrightarrow P$. The spaces C_1 and B have their universal coverings $\tilde{C}_1$ and $\tilde{B}$ in FS. Therefore, $H_i(\tilde{C}_1)$ and $H_i(\tilde{B})$ are finitely generated abelian groups for all i, and so are $\pi_i(\tilde{C}_1)$ and $\pi_i(\tilde{B})$ by the Serre-Hurewicz theorem [Hu, Chapter 10]. Hence, $\pi_i(\tilde{F})$ is a finitely generated abelian group for all i and so is $H_i(\tilde{F})$ by the Serre-Hurewicz theorem. Moreover, $\pi_1(F)$ has a subgroup of finite index which is a finitely generated abelian group. This implies that the space F in in FS by [Bi, Corollary 1.12]. As $P \in FS$, one deduces from Proposition (A.1) that $X \in FS$. []

REFERENCES

Only references **without** an asterisk contain results which are used in one of our proofs. Other are used for context discussions.

[Ba] BARGE J.
Dualité dans les revêtements galoisiens.
Invent. Math. 58 (1980) 101-106

[Bi] BIERI R.
Homological dimension of discrete groups.
Queen Mary College Lecture Notes, London (1976).

[Bn] BROWN K.
Homology criteria for finiteness.
Comm. Math. Helv. 50 (1975) 129-135

[Br1] BROWDER W.
Surgery on simply connected manifolds.
Springer-Verlag 1972.

[Br2] - Poincaré spaces, their normal fibrations and surgery. Inv. Math. 17, (1972) 191-202

* [Br3] - Torsion in H-spaces.
Annals of Math. 74 (1961) 24-51

* [Br4] - Homotopy type of differentiable manifolds.
Coll. on alg. topology, Aarhus (1962) 42-46

[BM] BARCUS W.D.-MEYER J.P.
The suspension of a loop space.
Amer. J. of Math. 80 (1958) 895-920

[CS1] CAPPELL S.-SHANESON J.
The codimension two problem and homology equivalent manifolds. Annals of Math. 99 (1974) 277-348

[CS2] CAPPELL S.-SHANESON J.
On four dimensional surgery and applications.
Comm. Math. Helv. 46 (1971) 500-528

[Do] DOLD A.
Chern classes in general homology.
Symposia Mathematica V (1969) 385-410

* [Di] DIEUDONNE J.
A history of algebraic and differential topology
Birkhauser (1989)

[EM] ECKMANN B.-MULLER
Poincaré duality groups of dimension two.
Comm. Math. Helv. 55 (1980) 510-520

* [EL] ECKMANN B. LINNELL P.
Poincaré duality groups of dimension two II.
Comm. Math. Helv. 58 (1983) 111-114

* [GS] GITTLER S.-STASHEFF J.
The first exotic class of BG
Topology 4 (1965) 257-266

[Ha1] HAUSMANN J-Cl.
Homological Surgery.
Annals of Math. 104 (1976) 573-584

* [Ha2] HAUSMANN J-Cl.
Fundamental group problems related to Poincaré duality. Canadian Math. Soc. Conf. Proc. 2, Part 2 (1982) 327-336

* [HH] HAUSMANN J-Cl. HAMBLETON I.
Acyclic maps and Poincaré spaces.
"Algebraic Topology, Aarhus 1982", Springer Lect. Notes 1051, 222-245

[HHu] HAUSMANN J-Cl. HUSEMOLLER D.
Acyclic maps. L'Enseignement Math. 25 (1979) 53-75

[HV] HAUSMANN J-Cl. VOGEL P.
The plus construction and lifting maps from manifolds.
Proc. Symp. in Pure Math. 32 (1978) 67-76

* [HV2] - Manifold categories of a Poincaré duality space
Differential Topology, (proc. Siegen Topol. Symp. 1987, Springer Lect. Notes 1350 (1989) 241-258

[Hr] HAEFLIGER A.
Differentiable embeddings of S^n in S^{n+q} for $q \geqslant 2$.
Annals of Math. 83 (1966) 402-436

* [Ho1] HODGSON J.P.E.
The Whitney technique for Poincaré complex embedings. Proc. AMS 35 (1972) 263-268

* [Ho2] - General position in the Poincaré duality category.
Invent. Math. 24 (1974) 311-314

* [Ho3] - Surgery on Poincaré complexes.
Trans. AMS 285 (1984) 685-701

[Hn] HUDSON J.E.P.
Piecewise linear topology. Benjamin Inc. 1969.

* [Ht] HUTT St.
Surgery on simply connected Poincaré spaces
Ph.D. Thesis, Univ. of Edimburgh (1989)

* [Jo1] JONES L.
Patch spaces : a geometric representation for Poincaré spaces. Annals of Math. 97 (1973) 306-343

* [Jo2] - Corrections for αpatch spacesα
Annals of Math. 102 (1975) 183-185

* [La] LATOUR L.
Chirurgie non-simplement connexe.
Sem Bourbaki 397 (June 1971) pp. 397.01-397.34

* [LLM] LANNES J.-LATOUR L.-MORLET C.
Géométrie des complexes de Poincaré et chirurgie.
Preprint, IHES.

* [Le0] LEVITT N.
Applications of engulfings.
Thesis, Princeton (1967)

* [Le1] - On the structure of Poincaré duality spaces.
Topology 7 (1968) 369-388

* [Le2] - Applications of Poincaré duality space to the topology of manifolds. Topology 11 (1972) 205-221

* [Le3] - Poincaré duality cobordism
Annals of Math. 96 (1972) 211-214

* [Le4] - A general position theorem in the Poincaré duality category.
Comm. Pure and Appl. Math. XXV (1972) 163-170

* [LR] LEVITT N.-RANICKI A.
Intrinsic Transversality structures
Pacific J. of Math. 129 (1987) 85-144

[MH] MILNOR J.-HUSEMOLLER D.
Symetric bilinear forms.
Springer-Verlag 1973.

[MM] MADSEN I.-MILGRAM J.
The classifying spaces for surgery and cobordisms of manifolds.
Ann. of Math. Studies, Princeton Univ. Press (1979)

* [Mc] MISHCHENKO
Homotopy invariants of non-simply-connected manifolds III. Izv. Akad. Nauk SSSR, ser. mat. 35 (1971) 1316-1355

[PR] PEDERSEN E.-RANICKI A.
Projective surgery theory.
Topology 19 (1980) 239-254

* [Qu] QUINN F.
Surgery on Poincaré and normal spaces
BAMS 78 (1972) 262-267

[Ra] RANICKI A.
Algebraic and geometric splitings of the K- and L-groups of polynomial extensions.
Proc. Poztnan Topology conf. (1986) Springer Lecture Notes 1217, 321-363

* [Ra2] - Exact sequences in the algebraic theory of surgery
Math. Notes, Princeton Univ. Press (1981)

* [Ra3] - The total surgery obstruction
Proc. Aarhus Conf. 1978, Springer Lect. Notes 763 (1979) 275-316

[Sh] SHANESON J.
Wall's surgery obstruction groups for $G \times \mathbb{Z}$.
Annals of Math. 90 (1969) 296-334

[Sp] SPIVAK N.
Spaces satisfying Poincaré duality
Topology 6 (1967) 77-102

[St] STEENROD N.
Homology with local coefficients.
Annals of Math. 44 (1943) 610-627

[Th] THOM R.
Quelques propriétés globales des variétés différentiables. Comm. Math. Helv. 28 (1954) 17-86

[Vo] VOGEL P.
On the obstruction group in homology surgery. Publ. Math. IHES 55 (1982) 165-206

[Wa1] WALL C.T.C.
Classification problems ... IV (Thickenings). Topology 5 (1966) 73-94

[Wa2] - Surgery on compact manifolds. Academic press (1970)

[Wa3] - Finiteness conditions for CW-complexes. Annals of Math. 81 (1965) 56-69

[Wa4] - Poincaré spaces I. Annals of Math. 86 (1967) 213-245.

[Wd] WALDHAUSEN F.
Algebraic K-theory of generalized free products I. Annals of Math. 108 (1978) 205-256.

[Wh] WHITEHEAD G.
Elements of homotopy Theory. Springer-Verlag, 1978.

[Wh2] WHITEHEAD J.H.C.
On simply-connected 4-dimensional polyhedra. Comm. Math. Helv. 22 (1949) 48-92.

INDEX